Chasing Infinity

Life's Début on Earth---
Original, Unborrowed, Underived

© Chuck Nelson

**Every living cell powering this elegant
Canada Goose flying machine, is infinitely more complex
than any mechanism ever designed by human genius.**

Cover

Rebecca Paschal's stunning cover design features
Mike Norton's glimpse of towering, scenic stone sentinels
highlighting western America's landscape.

Warren LeRoi Johns

Author of *Time Zero, Genesis File, Beyond Forever,
Ride to Glory,* and *Dateline Sunday, U.S.A.*

Chasing Infinity

Copyright © Warren LeRoi Johns, 2011

© Warren L. Johns

Publisher: www.GenesisFile.com
Printed in the United States of America
Lightning Source, LaVergne, TN

ISBN: 978-0-615-33381-6
Library of Congress Control Number: 2011920742

Recognizing Greatness

*"The great person, the great man,
is the miracle of history." ***

Moses, beneficiary of the finest education provided
ancient Egyptian royalty, is credited with authoring Genesis,
with its Creation account, and the four other books of the
Pentateuch. An intellectual giant with a commanding presence, Moses lived
without peer as lawgiver, courageous leader, and unselfish human dedicated
to serving God, the Creator of all things.

Peter fished in the Sea of Galilee until
Christ called him to be a *"fisher of men."* Enthusiastic, visionary, loyal,
courageous, and most of all a man of faith. Peter committed his life to
teaching the truth about God; the life, death and resurrection of Christ, the
life giver; and the miracle of creation.

On the road to Damascus, Saul of Tarsus, saw the light of truth.
As **Paul the Apostle,** he devoted his energies to presenting
the Bible's "good news" of faith, hope, and life eternal through Christ!
A legend of living courage and faith, Paul put his life on the line,
preaching the truth about God and His creation miracle.

* Carl Sandburg recognized Abraham Lincoln as a *"great person!"*
Genesis File honors five others who championed God's truth
and the miracle of His creation

A Fresh Look at the Evidence

Unlike superstitious imaginings suggesting first life organized itself from nothing, only to evolve without direction or purpose, eventually ending in the blackness of certain death, together, the Old and New Testaments of the Bible, combine to invest life with dimension.

Descriptive of an estimated one hundred miraculous events, the Bible itself unfolds a panorama of history's miracle in action. Sixty-six books, authored by inspired humans from Moses to John the Revelator, writing between 1400 BC and 100 AD, provide reason for living and direction for every *Homo sapiens* who has ever lived.

The "Big Picture" theme of Scripture begins with the miraculous origin of life on earth during the literal, seven-day creation week. The Bible's central figure, Christ, the Creator and God's only begotten son, whose life, death, and resurrection tie the ends of historical events together in a contiguous whole of meaning and hope.

The courageous Moses, an intellectual giant, believed to have authored the Genesis account of life's beginning, wrote from experience, perceptive insight and personal communication with God, Creator of all things!

Blessed with access to the sophisticated education reserved for ancient Egypt's royalty, his faith in the Creator was absolute given his cultural exposure to the eyewitness and written accounts of creation week and the global deluge handed down players like Adam, Seth, Noah, Shem, Abraham and Jacob, devout worshipers of God.

John the Beloved, walked by Christ's side during His ministry, and a witness to His crucifixion, penned the Book of Revelation, the last of the Bible's 66 books, with a look to the future and life eternal.

More than the Synod of Hippo's 66-book compilation of historical events, philosophy, poetry, and rules for better living, the Scriptures offer insight as to where we came from, what we are doing here and where we are going.

The Bible, the all-time best seller, speaks with inspired authority!

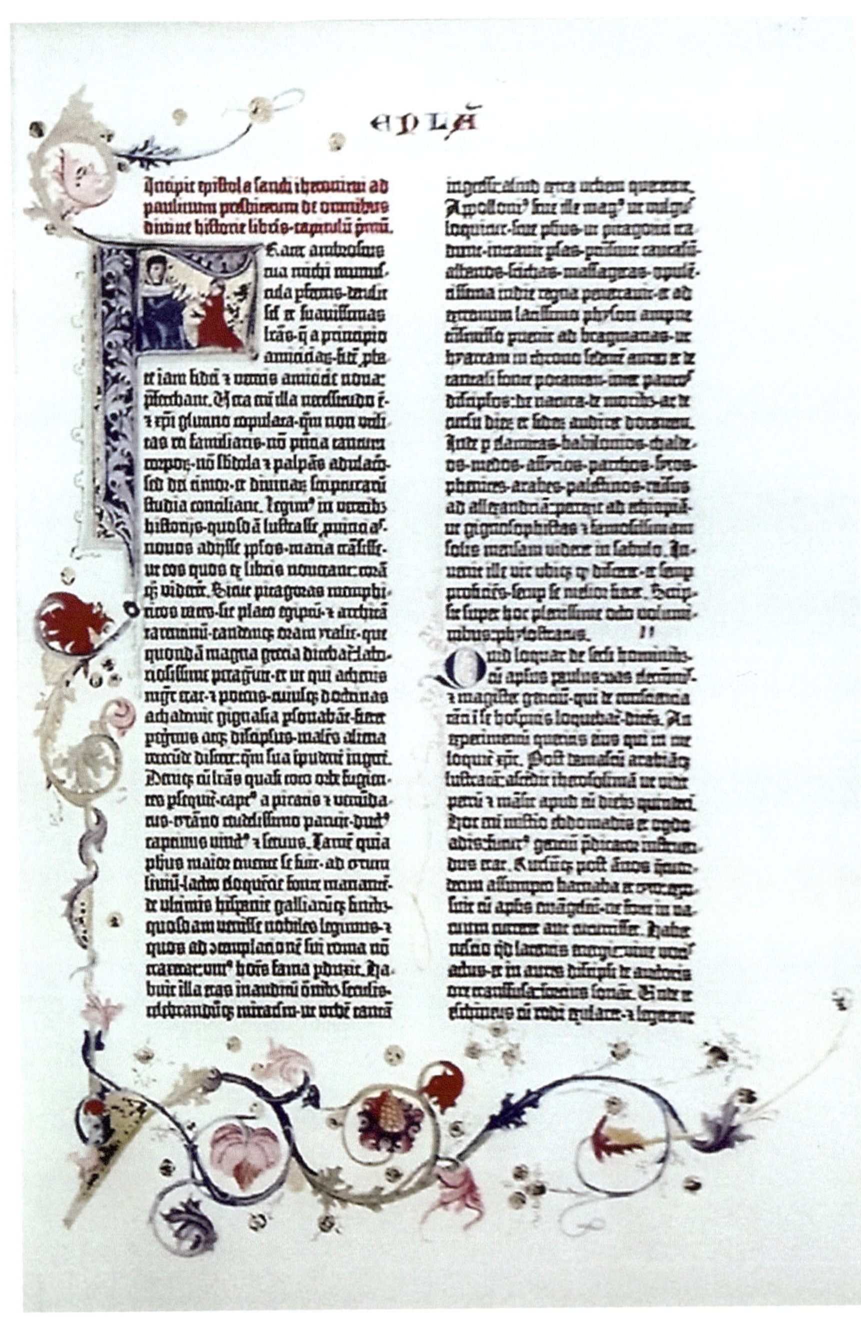

Johannes Gutenberg, pioneer book publisher, printed one-hundred eighty copies of the *Gutenberg Bible*. This Genesis 1 replica, complete with handcrafted illuminations, is a reproduction taken from one of the original copies still existing.

I

Life Conquers Death

Original, Unborrowed, Underrived

"King of Kings. Lord of Lords.
He shall reign forever and ever." [1]

George Frederic Handel

© egd

An artist's perception of "Christ the Redeemer," crowns
Rio de Janeiro's Corcovado. The outstretched arms invite
all people to accept the "Good News" from the cross.
"I am the way, the truth, and the life." [2]

"It is finished" were the last words heard from the anguished lips of a victim suffering excruciating pain, condemned to a death of cruel and unusual punishment by a tyrannical regime that had contempt for human life and equated justice with the power of the sword.

That Friday afternoon, early in the first century A.D., changed earth history forever and played to a larger audience in the cosmic theater of the universe.

Face bleeding from a crown of thorns jammed onto his head; taunted with a purple robe mocking kingly power; struck in the face; Christ endured a public flogging---all before being found guilty of anything.

His crime: three years serving a suppressed populace with a message of healing, hope, and happiness. After questioning Jesus of Nazareth, the pathetic Pilate declared, *"I find no basis for a charge against Him."* [3]

Still, the unruly mob ruled the courtyard chaos!

In a dark-of-night trial, an innocent thirty-three-year-old was sent to His death by Pilate, a weak-kneed political hack, intimidated by frenzied fanatics that thumbed collective noses at legal due process.

Christ surrendered His life on the Cross, voluntarily!

Execution by death on a cross represented Roman "justice" at its worst. A jeering crowd trailed His steps as Roman soldiers placed a cross on His shoulders and led the way to Jerusalem's *hill of a skull*, the execution site.

© MaxFX

Torture was added to the mix by nailing the hands and feet to rough, tree-cut lumber, and then thrusting the upright device into the ground to maximize hurt. Left to writhe in pain for hours, or even days, until the victim took a final gasp of breath.

Cruel and unusual punishment even for the most despicable criminal.

In this case, the apparent triumph of evil, changed world history. Christ the Creator, God's son in human form, lived and died to demonstrate to the universe that Lucifer's hatred, would kill even God, if he could.

The Scriptural lexicon defines evil as sin.

Lucifer the "light bearer," a high ranking created being, overcome by pride and jealousy, challenged God's authority, justice, and love, and stirred doubts, resentment and rebellion in the hearts of other angels.

Sin deceives the *"very elect,"* delivering death.

Had God suppressed the rebellion by destroying Satan, the act would appear to justify the charge of tyranny with the result that other created beings, fearful for their own lives, would be inclined to serve God from fear rather than free choice inspired by love.

Instead, the adversary of all things good was cast from God's heavenly presence and confined to earth, center stage for a raging spiritual battle demonstrating the tragic consequences of evil destined to be witnessed by a universal audience.

The Genesis narrative describes the cunning deception of Satan, the fallen Lucifer, now God's adversary determined to mar a perfect creation and to recruit humans to join his rebellion.

Sin's ripple effect disfigures perfection and devours life---casualties of the spiritual war pitting good versus evil.

Despite imposition of death's curse imposed on mankind for disobedience and failure to trust, God offered a *"way of escape"* through the plan of redemption demonstrating His love and fairness.

Inescapably, the plan imposed death as sin's wage.

God's son paid that cost for all sinners. Had God the Father died to demonstrate His love for humans, Satan would have won. So God sent His *"only begotten son,"* whose life and ministry demonstrated ultimate love.

Christ's death on the cross, revealed Satan's hatred and God's unbounded love and commitment to all humans choosing to believe. God's Spirit, with His power for forgiveness and love, justifies the life of any human choosing to believe.

Until Christ visited earth in person, the image of original power and created perfection blurred, lost in trivia and a cosmic jumble. Jesus, of

Nazareth, the greatest life ever lived on Planet Earth, arrived with less than auspicious human credentials.

He was born in a barn, burdened by rumors of suspect parentage.

Surrounded by poverty, Jesus lived in a culture gasping for survival under the brutal heel of Rome's ruthless "iron empire."

Robust, bronzed and brawny, He served an apprenticeship in Joseph's carpenter shop. Less than a bed-of-roses environment, corruption shrouded the streets and alleys of his home village.

Aware of the town's notoriety, cynics raised eyebrows, asking rhetorically, *"Can any good thing come out of Nazaret©*

© Nathan Greene

Christ never had a bank account, owned a home, wrote a book, held political office, or commanded a military force.

Exuding charismatic personal charm, all it took were the words, "follow me" for twelve young men to join his unique ministry of hope, goodness and the promise of abundant living today and a forever life to come.

His disciples followed with no promise of pay or explanation of just where He would lead. No one asked about perks. They accepted his invitation in faith---simply because he said, *"Come, follow me and I will make you fishers of men."*

Some misunderstood the invite, hoping the Messiah would deliver subjugated citizens from the political heel of the Roman oppressor.

Despite lack of academic credentials, He taught thousands, captivating minds with a message of love, forgiveness, and a formula for peace. No microphone boosted the sound of His words.

He lived and led by example from the geographic crossroads of world commerce. Without access to mass media, news of His words and deeds traveled with lightning speed. No media blitz publicized his appearances. Rumors that Jesus worked miracles, focused attention to His ministry.

He didn't sermonize from the altar of a cathedral or a pulpit in an arena seating thousands.

Still, thousands followed His footsteps, captivated by a message that echoed across cultures and two millenniums into the future.

When a wedding celebration ran out of wine, He transformed water into prime vintage by His word alone. Powered by His blessing, five loaves of bread and two fish satisfied the hunger of a crowd of 5,000.

With the curse of incurable leprosy running rampant, hearts beat faster and eyebrows raised with news He restored vibrant health to ten lepers.

Responding to His command, an infirm man *"took up his bed and walked."* Restoring sight to a blind man startled observers who witnessed the incredible scene.

The countryside pondered His wisdom and responded to His leadership. Looking for clues to empowerment, hundreds followed Him to a natural, outdoor setting on the slope of a hillside. Tickets were not required.

He spoke with authority.

But to the surprise and chagrin of some who came prepared to take up arms against the Roman occupier, He offered words to inspire the human spirit instead, blazing a trail to inner peace and future promise.

He predicted the *"meek"* would *"inherit the earth."*

"Blessed are the merciful, for they shall be shown mercy. Blessed are the peacemakers, for they shall be called the children of God...." [4]

Contrary to epic notions of heroic history built around military conquest with its harvest of death and destruction, Christ created original life and preached peace and the sanctity of life.

He urged listeners to seek right; to be pure in heart; to love enemies; to forgive up to seven-times-seventy; and to reconcile with adversaries. As the antithesis of darkness, he reminded followers:

"You are the light of the world...Let your light shine before men that they may see your good deeds and praise your Father in heaven." [5]

The words sounded a call to spiritual arms, not military action!

This formula for a better life opened the door to personal peace and abundant living. It came with the gift of a Divine Comforter, empowering believers to transform to a new and improved version of themselves.

Although Christ committed no crime, His human body died the death of a criminal, an innocent victim of man's inhumanity to man.

Within minutes after His death, He was buried in haste. No funeral or memorial service honored His life. Nor was He recognized with public burial rites featuring eloquent rhetoric.

Joseph of Arimathaea volunteered his own family's tomb, chiseled from hillside rock, as a final resting place. But the "rest" was not to be "final."

After the Sabbath, early the first day of the week, the seal of the tomb was broken, the stone entry door rolled away, and Christ rose, forever victorious over death.

The crucifixion events crowned by the resurrection miracle, described in the writings of Mathew, Mark, Luke and John, are preserved for all to read in the Biblical cannon composed of Old and New Testaments compiled by devout Christian leaders late in the fourth century A.D.

Christ's immaculately conceived birth, flawless life in the midst of an evil environment, cruel crucifixion and unselfish death, crowned by the resurrection miracle, is unmatched in other belief systems.

Ironically, some of the same evolutionists who doubt Christ's resurrection, somehow manage to believe life on earth created itself from non-living matter, accidentally, by spontaneous generation.

Christ, the Creator and source of all life on earth, exemplified God's absolute justice and patient love for humanity, demonstrating unlimited Divine power---not just to hang the sun and the moon but also to conquer death's curse.

The Scriptural "Big Picture" narrative, written over a 1500 year time span, begins with Moses' Genesis account of the creation miracle and concludes with John's Revelation describing Christ's return to earth bringing life eternal to all who believe.

Christ's ministry revolutionized history!

Christ's unassailable victory over death, as reported and recorded by living witnesses, left God's adversary no option but to surrender to failure or to attempt to rewrite history by launching a frontal attack.

Having deceived Adam and Eve in the Garden of Eden, the diabolical prevaricator initiated the same deceptive strategy to discredit God by depicting the Scriptural narrative as myth and life on earth as nothing more than a mindless, evolving accident.

Conjecturing that life originated millions of years in the ancient past in some unidentified *"warm little pond"* echoes the fiction. Some of earth's brightest minds swallow the insidious line!

Life on earth didn't start accidentally by spontaneous generation in some rocky, *"warm little pond!"* The miracle of life is a gift from the Rock of Ages.

II

Fact-free "Science"

Phantom Chasing Imaginings

"I believe that one day the Darwinian myth
will be ranked the greatest deceit
in the history of science." [1]

Søren Løvtrup

© janprchal

Ocean-going whales are out of sync with evolution's sea-to-land
scenario. Darwin's imaginary solution: *"monstrous"*
whales might have evolved from bears.

Charles Darwin rejected the Genesis narrative of life's origin.

Undeterred by lack of verifiable scientific evidence explaining the origin of first life and relying on his own "warm little pond" scenario, Darwin plunged ahead with his grand scheme alleging every diversified plant and animal life form shared common ancestry with that yet to be identified, original, simple cell.

In his initial burst of fervor to illustrate this extravagant notion, Darwin postulated that given time, a bear might evolve into a marine mammal *"as monstrous as a whale."*

This land animal-to-ocean-going-whale fiction appeared as a figment of Darwin's fertile imagination in the first edition of his *Origin of Species*.

"In North America the black bear was seen by Hearne swimming for hours with widely open mouth, thus catching, like a whale, insects in the water.

"Even in so extreme a case as this, if the supply of insects were constant, and if better adapted competitors did not already exist in the country, I can see no difficulty in a race of bears being rendered, by natural selection, more and more aquatic in their structure and habits, with larger and larger mouths, till a creature was produced as monstrous as a whale." [2]

This abstract, thumb-your-nose rationale conflicts head-on with the Biblical reference that *"God created great whales."* [3]

The nineteenth century naturalist never enjoyed the privilege of donning hip-waders and fraternizing with beluga whales in a 55° Chicago water tank designed to entertain the human species.

These aquatic monsters weigh-in at a ton or more, and in captivity open their mouths to be fed, have their tongues petted, and to nuzzle noses with human kind who fork over $200 for the privilege.

Darwin's whale-of-a-tale scenario imagining a bear evolving into a sea-going whale exposed the glaring inconsistency of a flawed idea!

© Audrey Snider-Bell

No question, bears wade in water to catch fish. Other than that, even Darwin sensed the impossible bear-to-whale stretch and deleted the reference in *Origin* editions after 1859.

Does anyone seriously believe the beluga that bedazzles humans, evolved coincidentally and gradually from some ferocious, land-based, hairy-bear ancestor?

Stuck with evolution's fact-free fantasy from fish-to amphibians-to-reptiles-to-mammals, the appropriate place of the whale in the sequence, posed a conundrum.

If the warm-blooded whale, a mammal, descended from an egg-hatched, cold-blooded reptile, the chain of organic life transitions stood in jeopardy, seriously out-of-kilter..

A winter hibernating, fury, land-based mammal with paws and claws, transiting to a sleek denizen of the deep-blue-sea *"as monstrous as a whale"* ignores logic and lacks supporting evidence.

Suggesting *Ambulocetus* is a "walking whale," an extinct, 11 foot, four-legged, land-based critter, lacking a blowhole, flippers and missing a whale's rudder-like tail, matches Darwin's bear-to-whale absurdity.

Why would descendants of a self-sufficient, land based bear, seek refuge in the ocean's salt-water depths in order to survive?

Had the imagined bear-to-whale evolution actually occurred, why do bears roam the landscape after Darwin predicted *"We may safely infer that not one living species will transmit its unaltered likeness to a distant futurity."* [4]

Understandably, Darwin himself presumably recognized the preposterous absurdity by abandoning the conjecture in all post-1859 editions of *The Origin of Species.*

Any bear attempting transit to whale status by natural selection or otherwise would be destined for a watery grave---as helpless as a car without wheels or as impractical as a desk-top printer without an ink cartridge.

While this ambitious bear-to-whale postulation never made it past 1859's first edition of *The Origin of Species,* Darwin later underscored his vision of the make-believe with a conjectured *"molecule-to-man"* scenario resulting from a series of un-designed natural accidents.

Charles Darwin, living in a cocoon of privilege with access to a family fortune, didn't have to earn a living with his hands.

The well-to-do Darwin family supported his university studies while he explored possible careers in medicine and in the ministry. The family connection earned him crew membership as a naturalist on the HMS Beagle. Later, married to an heir to the Wedgewood fortune, he settled in a country estate near London.

Darwin expressed his thoughts during a time when the sun allegedly never set on the British flag and empire ruled politics. While tin mines were closing in Cornwall and men were dying in the Crimean War, he enjoyed the comfort of an estate managed by an eight-person servant staff.

Thriving in his personal cocoon of privilege, he made no apology for inherited wealth and the social perks that came with the turf.

"The presence of…well-instructed men, who have not to labour [sic] for their daily bread, is important to a degree which cannot be overestimated; as all high intellectual work is carried on by them, and on such work material progress of all kinds mainly depends." [5]

Darwin entered the world in 1809, near the peak of a power pyramid at a moment when "Rule Britannica" symbolized empire. British flagged slave ships plied the Atlantic and wealth was built on the backs of foreign nationals and an exploited home-grown labor force that lived at the whim of the owners of the mills and mines

In a coincidence of social irony, Abraham Lincoln, the author of the "Emancipation Proclamation," shared a February 12, 1809, birth date with Charles Darwin, propagator of racial supremacy.

His words indicate awareness that he lived near the apex of the not-so-subtle class structure that saddled the system.

"Without the accumulation of capital the arts could not progress; and it is chiefly through their power that the civilized races have extended, and are now everywhere extending, their range, so as to take the place of the lower races." [5]

He appears to have assessed accumulation of "capital" a valid measure of human worth. With personal vision warped by nineteenth-century superstitions, Darwin fretted that vaccination would lead *"to the degeneration of a domestic race."*

Darwin's survival of the fittest theme inspired his taking an ill-conceived shot (pun intended) at vaccination, believing it *"preserved thousands, who from a weak constitution would formerly have succumbed to small-pox…the weak members of civilized societies propagate their kind…this must be highly injurious to the race of man…a want of care, or care wrongly directed, leads to the degeneration of a domestic race."* [6]

Regardless, in the classic tradition of his English heritage, he ventured to maintain a stiff upper lip while pledging to *"bear without complaining the undoubtedly bad effects of the weak surviving and propagating their kind."* [5]

Given the flourishing slave trade sanctioned by his society, racism tainted his public pronouncements. Not surprisingly, he awarded pinnacle status to his home turf.

"Various races differ much from each other…the capacity of the lungs, the form and capacity of the skull…in their intellectual, faculties." [7] *"The western nations of Europe…immeasurably surpass their former savage progenitors and stand at the summit of civilization…"* [8]

He predicted, thanks to evolution's touted *"progress toward perfection,"* the time would come *"…not very far distant, as measured by centuries, the civilised [civilized] races of man will almost certainly exterminate and replace throughout the world the savage races."* [9]

England's Queen Victoria, a Darwin contemporary, occupied the British throne for 63 years beginning in 1837. Without a tip of the hat to the lady monarch, Darwin revealed a gender bias suggesting that after moving past monkey ancestry, *"Man has ultimately become superior to woman."* [10]

To the Queen's credit, Darwin wasn't banished to the Tower of London for saluting male chauvinism! Undeterred, the naturalist elaborated further, exposing more than a tinge of arrogance with a hint of ignorance! His assumptive pronouncements left no equivocation.

"If two lists were made of the most eminent men and women in poetry, painting, sculpture, music,---comprising composition and performance, history, science, and philosophy, with half-a-dozen names under each subject, the two lists would not bear comparison.

Undeterred by lack of science credentials earned in a university study program, Charles took his cue and shaped his views from personal research and the leanings of his Grandfather Erasmus.

Erasmus Darwin speculated that *"all warm-blooded animals have arisen from one living filament"* that could continue *"to improve by its own inherent activity"* with the improvements passed along to descendants. [13]

In 1837, just back from a five-year odyssey aboard the HMS Beagle, the 28-year-old Charles sketched a "tree of life" depicting branches of new and different life forms sprouting from that single living cell with its unknown roots of chance origin.

Twenty-two years later, his *"I think"* tree blossomed into the ponderous *Origin of Species* treatise conjecturing relentless shifting of unstable life forms, in constant transit from the simple to the complex, en route to an unpredictable "biologic transit stop."

While blissfully sweeping under the rug the *"far higher problem"* as to just how life sprung accidentally from non-living matter, he embraced *"natural selection"* as the primary means for culling out and preserving the fittest in evolution's cavalcade of change.

With phraseology more than faintly reminiscent of Grandfather Erasmus' musings, Darwin's grand scheme imagined a taxonomic tree, with new and different branches sprouting from the accrual of multiple miniscule increments of genetic change, transiting to new and distinctly different organisms.

Genetically preposterous bear-to-whale mismatches exposes evolution's make-believe for the extravagant musing that it is while doing nothing to validate Darwin's vision of a chain of organic life transiting upward from that first living cell *species*, and on to *genus*, to *family*, to *order*, to *class*, to *phylum*, and eventually to *kingdom*.

Charles Robert Darwin
(1809-1882)

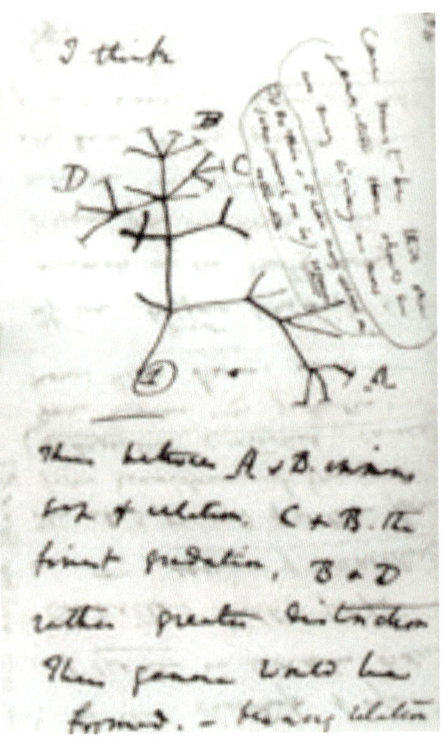

A 28-year-old Darwin
designed this "I Think" tree
in 1837.

Library of Congress

American Botanist, Asa Gray,
a Darwin correspondent in
the United States.
(1810-1888)

Alfred Russell Wallace, British
evolutionist, shared many
of Darwin's ideas.
(1823-1913)

The inconvenient truth stands resolute: no evidence exists suggesting real change within a genome has ever jumped genetic limits, by evolving radically different descendants that creep gradually up the *family-order-class-phylum-kingdom* taxonomic ladder!

Consider Darwin's words!

Natural selection can only take *"advantage of slight successive variations; she can never take a great and sudden leap, but must advance by short and sure, though slow, steps."* [14]

Those *"slight successive variations"* supposedly result in *"the production of new forms"* which will cause *"the extinction of about the same number of old forms."* [15]

Has anyone seen *"slight successive variations"* produce a partially evolved organ such as an eye, lung, heart or brain?

In the pointed vernacular of down-home Texans, "You can put your boots in the oven but that can't make them biscuits."

Natural selection is a genuine fact of nature!

But don't hold your breath expecting Darwin style evolutionary change if *"natural selection"* relies either on acquired physical characteristics or mutations to jump genetic barriers inherent in every genomic kind.

In reality, a female's *sexual selection*, expressing preference for a mate, occurs throughout animal kingdom kinds. Males attract a female's attention by performing exotic dances, feats of strength and issuing "romantic" sounds, unique to the species.

It's the female's prerogative to select the winner!

Rather than evolving some new and different organism, her selection works to preserve, protect and maximize the genetic quality of her own species---the precise opposite of the Darwinian postulate.

Working at cross-purposes to evolution theory, *"natural selection"* acts counter to the evolution equation by screening out harmful mutations.

"Natural selection can serve only to 'weed out' those mutations that are harmful, at best preserving the 'status quo.' "[16] It *"…can act only on those biologic properties that already exist; it cannot create properties in order to meet adaptational needs."*[17]

"No one has ever produced a species by mechanisms of natural selection."[18]

Sexual selection as an obvious demonstration of natural selection? Of course!

Natural selection as a demonstration of evolution in action? No way!

Darwin reasoned that a simple form of first life might have emerged, accidentally, from some vaguely defined, *"warm little pond."*[19]

Then given enough time, that simple life supposedly evolved into *"an aquatic animal…with the two sexes united in the same individual;"* then on to *"some fish-like animal"* ancestral to a *"reptile-like or some amphibian-like creature;"* ultimately morphing into an *"ancient marsupial"* and *"higher mammals."*[20]

"All the higher mammals are probably derived from an ancient marsupial, and this through a long line of diversified forms, either from some reptile-like or some amphibian-like creature, and this again from some fish-like animal."[20]

Darwin assured skeptics that *"Man is descended from a hairy quadruped, furnished with a tail and pointed ears, probably arboreal in its habits…"*[19] that *"early progenitors of man were no doubt once covered with hair, both sexes having beards; their ears were pointed and capable of movement; and their bodies were provided with a tail…"*[21]

He alleged humans could trace their ancestry to *"Old World monkeys."*[22] His ponderous manuscripts appear less clear as to just where the plant kingdom fit into evolution's master scheme.

The master of the Downe estate, never retracted the core of his unproven imaginings, but he did express misgivings to trusted confidents!

Without bowing to the Genesis account of life's origin but lacking any rational alternative, Darwin conceded "original life" may have been *"originally breathed by the Creator into a few forms or into one."* [23]

Michael Denton, a biochemist writing late in the twentieth century, saw *"imagination"* as an inadequate substitute for *"evidence."*

"If anyone was chasing a phantom or retreating from empiricism it was surely Darwin, who himself freely admitted that he had absolutely no hard empirical evidence that any of the major evolutionary transformations he proposed had ever actually occurred.

"It was Darwin, the evolutionist, who admitted in a letter to Asa Gray, that one's 'imagination must fill up the very wide blanks.' " [24]

In a burst of candor, Darwin recognized the feeble fabric of his idea, acknowledging, *"I am conscious that my speculations run beyond the bounds of true science."* [25] while admitting also his theory was *"a mere rag of an hypothesis with as many flaw[s] and holes as sound parts."* [26]

Before his 1882 death, the English naturalist verbalized uneasy equivocation in a whiff of prescience, confessing doubts about his thesis. He fretted he may *"…have devoted my life to a phantasy [sic]."* [27]

His concerns were justified.

To qualify as science, *"Explanations of large classes of phenomena must make testable predictions and be falsifiable…there must be a way to make an observation that could disprove the explanation."* [28]

Labeling a sow's ear doesn't make it a silk purse. Hyping evolution's abstract assumptions doesn't qualify unproven imaginings as science.

Facts have a way of fading in the hands of artful spinmeisters capable of whitewashing assumptions, affixing catchy labels, and wrapping conjecture in cosmetic mantles of pseudo-authenticity.

Evolution Darwin style doesn't happen…it never did, its not happening now, it never will!

To paraphrase Winston Churchill, building an idea on a bubble *"is like a man standing in a bucket trying to lift himself up by the handle."*

III

Cosmic Convergence

Life-friendly Ecosystem

"You may find it hard to believe God could make everything from nothing. But the alternative is that nothing turned itself into everything." [1]

Mark Cahill

Hubblesite.org

Eagle Nebula's awesome niche in the infinity of space!

Earth's life-friendly ecosystem could not have created itself.

Against all odds, some twenty-first century theorists came to the rescue, proposing an unusual "good" news/"bad" news scenario!

The "good" news: The unconfirmed report that the mystery of the origin of first life on Planet Earth has been solved?

The "bad" news: The suggestion our ultimate ancestor may well have been a *"biochemical moron?"*

The report starts simply enough, announcing that *"…many scientists believe that viruses evolved very early on, possibly even earlier than everything else.*

"If so, they are not merely some ornamentation on the tree of life but rather may compose its very roots." [3]

Our first ancestor was a virus parasite? Could a parasite virus live without access to a prior life form to latch onto?

But wait! There's more!

"We humans…are nobody's great idea; we are the fortunate mistakes of countless biochemical morons.

"That's evolution. It is humbling but somehow comforting." [3]

This is *"comforting?"* Finding some great-grand-pappy virus smirking at us smugly from the pages of the family's ancestry album?

Taking a swipe at Intelligent Design theory, the report asserts, *"…the viruses appear to present a creation story of their own: a stirring, topsy-turvy, and decidedly unintelligent design where life arose more by reckless accident than original intent, through an accumulation of genetic accounting errors committed by hordes of mindless microscopic replication machines."* [3]

© Paul Prescott

The moon's earth orbit can be calculated hundreds of years, past or future! Evolution can't explain how life "evolved" on our Blue Planet but bypassed the sterile moon!

"Unintelligent" may be the understatement of the millenium!

Throw in *"reckless accident…mindless…genetic accounting errors… mistakes…"* and, of course, those *"countless biochemical morons"* and the imagined trip to antiquity begins to resemble a journey to la-la land.

While the verbiage may inspire a field day of punditry, the idea carries a serious side. Recognizing viruses to be older and more complex than once believed and that possibly they may compose the *"very roots"* of the *"tree of life"* represents reckless, eye-rolling poppycock!

Where's the substantiating evidence? And where's any verifiable explanation as to the source of genetic information for the virus?

Formulation of a living cell, capable of reproducing itself, has never been created in the laboratory much less accidentally generated spontaneously in nature.

Not even a parasite virus, incapable of independent living on its own without latching on to an already living host!

How can "something" explode from "nothing?"

What caused "nothing" to explode into "everything?"

If "nothing" was all that existed, what could go "bang?"

Could there be absolutely "nothing"---no energy, no matter, no light, not even empty space in a universe without boundaries?

Do the laws of physics support the notion that every piece of matter floating in a 14-billion light year observable universe could have been compacted in a dot no bigger than the period closing a sentence?

If energy and matter can neither be created nor destroyed, then where did the energy and matter compressed within that miniscule dot originate before the supposed inflation from the big boom-boom? And how can a destructive force create a cosmic universe of something out of nothing?

Verifiable answers to these queries ride waves of silence!

For the cosmos to have created itself as conjectured, the basic laws of physics would have to be suspended, modified or abandoned!

"…When you squeeze the entire universe into an infinitesimally small, but stupendously dense package, at a certain point, our laws of physics simply break down. They just don't make sense anymore." [4]

Astrophysicist Peter Coles from England's University of Nottingham throws cold water on the not-so-hot idea.

"There is little direct evidence that inflation actually took place…It is a beautiful idea that fits snugly with standard cosmology…but that doesn't necessarily make it true.

"We don't know for sure if inflation happened…In a way we are still as confused as ever about how the universe began." [5]

So much for the alleged "Big Bang!"

Meanwhile, back on Planet Earth. Why life here and not on the moon?

Big bangers look to an explosive force, supposedly occurring several billions of years before the present, as having self-created multi-trillions of bits and pieces of matter, which formed orbiting spheres, parading in space in some good-luck sequence of universal order.

Other academics attribute the harnessing of all the factors essential to create a life-friendly environment to a Master Designer!

Thanks to predictable order in space, mathematical measurements plot the time and place of cosmic orbits with uncanny precision, whether the time dimension dips millenniums into the past or extends outside the reach of an uncharted future.

How did a cosmic explosion create the mathematical balance evident in galaxies---from nothingness?

Suspended in space, without cables or foundations, earth moves in three directions simultaneously---spinning on its axis; orbiting the sun; and floating in sync with the other components of the Solar System within the Milky Way galaxy.

Nothing random about this Solar System of planets spinning about the sun; nothing chaotic; not the product of unpredictable chance!

The most brilliant minds can't duplicate this cosmic balancing act. Keeping free-floating spheres in perpetual motion, floating in repetitious orbits, without strings, confounds.

That's just the beginning of flabbergasting realities!

The length of a day on our Blue Planet is 24 hours with 365.256 days to a year while a Venus day runs 116.75 days and 1.92462 days per year. While every planet in the Solar System moves at its own speed, marching to its own drummer, the Solar System displays a precise symphony of perpetual, mathematical balance.

Even with the planetary data in hand, human "genius" that presumes life originated by accident, are congenitally incapable to creating a miniature Solar System model that will float and spin perpetually without props.

Explosions observed on earth rip matter apart, leaving fragments of ruble, strewn helter-skelter in disorganized trash heaps.

Where is evidence a cosmic explosion created order out of chaos?

Planetary orbits display variances of distance and shapes with tendencies ranging from the round to the elliptical. Axis tilt angles are unique for each planet. Most rotate counter clockwise to the east while Venus, Uranus and Pluto rotate clockwise to the west.

So what is correct? Does this sound like fall-out from an explosion?

Hubblesite.org

**Earth's Solar System orbits within the Milky Way Galaxy
that spreads its 100,000 light year diameter across the heavens.
Earth's sun orbits 25,000 light years from the galaxy's center.**

Can a cosmic explosion create order from chaos or does it destroy?

Can a theoretical explosion explain the origin of environmental conditions essential for life's genesis?

In our Solar System, earth alone exhibits a confluence of environmental factors essential to sustain organic life. Plants and animals can't survive except as inter-dependent components of an ecological package deal.

Recent estimates account for more than 1,000 planets in the cosmos--- and still counting. With billions of stars in space, it should be no surprise that life comparable or superior to humans, likely exists on planets with their own co-dependent ecosystems.

"No matter how large the environment one considers, life cannot have had a random beginning...[4] *Heavy* doses of intellectual flim flam can't disguise reality!

"There are about two thousand enzymes, and the chance of obtaining them all in a random trial is only one part in $(10^{30})^{2000} = 10^{40,000}$*, an outrageously small probability that could not be faced even if the whole universe consisted of organic soup...*

"If one is not prejudiced either by social beliefs or by a scientific training into the conviction that life originated on the Earth, this simple calculation wipes the idea entirely out of court..." [6]

"...The difficulties in producing a protein from the mythical prebiotic soup are very large, but more difficult still is the probability of random processes producing the simplest living cell..." [7]

Think about it!

"The concept that all the parts of the first living thing preexisted, and its formation was simply a matter of spontaneous generation there from is mathematical absurdity, not probability. All present approaches to the problem of the origin of life are either irrelevant or lead to a blind alley." [8]

Bradley and Thaxton reason *"...even assuming that all the carbon on earth existed in the form of amino acids and react at the greatest possible rate of* $10^{12}/s$ *for one billion years...the mathematically impossible probability for the formation of one functional protein would be* $\sim 10^{65}$*."* [9]

Sir Fred Hoyle calculated *"the likelihood of even one very simple enzyme arising at the right time in the right place was only one chance in* 10^{20} *or 1 in 100,000,000,000,000,000,000...that about 2,000 enzymes were needed with each one performing a specific task to form a single bacterium like E. coli."* [10]

Hoyle, with his colleague Chandra Wickersham, estimated *"the probability of all of these different enzymes forming in one place at one time to produce a single bacterium, at 1 in $10^{40,000}$."*

"This number is so vast that it amounts to total impossibility." [10]

While sticking to his life-by-accident postulate,

Darwin tossed a sop to skeptics, conceding life had been

"originally breathed by the Creator into a few forms or into one..." [11]

Hardly an endorsement of the Biblical creation week, this allusion from the last paragraph of *Origin* may have appeased some Christians looking for compromise, while giving root to his *"one"* or a *"few"* simple life forms supposedly growing a *"tree of life."*

"It is often said that all the conditions for the first production of a living organism are now present, which could ever have been present.

"But if (and oh! what a big if!) we could conceive in some warm little pond, with all sorts of ammonia and phosphoric salts, light, heat, electricity, etc., present, that a protein compound was chemically formed ready to undergo still more complex changes, at the present day would be instantly devoured or absorbed, which would not have been the case before living creatures were formed." [12]

Without that first-ever living cell making its début, Darwin's *Origin of Species* dream of evolving branches on a grandiose *"tree of life"* would be stranded rootless with no chance to grow---even in a billion years!

Darwin's biological evolution hypothesis can't fly without the initial launch from that magical *"warn little pond"* scenario which posits the alleged action of energy sources forming organic compounds in the atmosphere *"washed down by rain and accumulated in the primitive oceans until they reached the consistency of a hot dilute soup. According to this model, life appeared from the chemical reactions and transformations that took place in this prebiotic soup."* [13]

Big problem here: evidence of the magic elixir is a no-show!

"Prebiotic chemical soup, presumably a worldwide phenomenon, left no known trace in the geological record." [14]

**The Orion constellation extends 24 light years across and
approximately 1344 light years distant from earth.
Orion may have looked like this circa 656 A.D.**

So-called *"dawn rocks"* from Western Greenland, conventionally dated at 3.9 billion years before the present and reputedly the oldest known dated rocks on the planet, show nothing resembling prebiotic soup.

"Rocks of great antiquity have been examined…and in none of them has any trace of abiotically produced organic compounds been found…

"Considering the way the prebiotic soup is referred to in so many discussions of the origin of life as an already established reality, it comes as something of a shock to realize that there is absolutely no positive evidence for its existence." [15]

Hubert Yockey dismisses *"primeval soup"* as a non-event.

"The origin of life by chance in a primeval soup is impossible in probability in the same way that a perpetual motion machine is impossible in probability." [16]

Abiogenesis didn't happen but should be viewed as *"just a relic of the cosmology of the time it was invented…*

"There is no evidence that a 'hot dilute soup' ever existed. In spite of this fact, adherents of this paradigm think it ought to have existed for philosophical or ideological

reasons…Scientists are divided into segregated schools that do not even agree on the standards of scientific inquiry…" [17]

With respect to the *"prebiotic soup theory of the origin of life…objective scientific principle of a search for the truth is replaced by the subjective aesthetic principle of a well-constructed story."* [17]

The never-happened spontaneous generation comes burdened in the collateral fiction of the never-was Urschleim, an imaginary slime-like material existing at ocean depths, supposedly functioning as the nursery for first life in a *"self-origination"* format.

"In science one must follow the results of experiments and mathematics and not one's faith, religion, philosophy or ideology. The primeval soup is unobservable since, by the paradigm it was destroyed by the organisms from which it presumably emerged." [18]

Darwin contemporary, Sir William Dawson, didn't think much of evolution theory!

He labeled it *"one of the strangest phenomena of humanity…a system destitute of any shadow of proof, and supported merely by vague analogies and figures of speech….*

"Now no one pretends that they rest on facts actually observed…Let the reader take up either of Darwin's great books, or Spencer's 'Biology,' and merely ask himself as he reads each paragraph, 'What is assumed here and what is proved?' and he will find the whole fabric melt away like a vision…

"Evolution as an hypothesis has no basis in experience or in scientific fact, and… its imagined series of transmutations has breaks which cannot be filled." [19]

Some evolutionists, sweep the origin issue under the rug by suggesting life survived the rigors of outer space travel and rode to earth after originating somewhere else in the universe following a cosmic Big Bang. But this cop-out, side steps the origins issue.

The likelihood of natural forces delivering the ingredients essential to sustain life, simultaneously in one place, at one instant in time, and without intelligent input is as preposterous as the discredited notion that spontaneous generation might create life from non-living matter.

Microscopic particles, one ten-millionth the mass of an electron, are believed to be the most miniscule form of matter presently known to man.

The atom, with its positively charged nucleus encircled by an array of electrons, ranks as the smallest unit of the elements charted in the periodic table displayed in High Schools.

Grc.NASA.gov

**Human minds harness natural laws to explore space.
The verifiable Law of Gravity represents genuine science.**

Inorganic molecular matter can be built from a mix of these elements in carefully measured, inorganic chemical recipes producing results capable of replication ad infinitum.

Molecular bonding, essential to life, requires the presence of no less than 40 different elements. Success is contingent upon the force of electromagnetism functioning within a balanced electron-to-proton mass ratio.

Starting with a formulated mix of oxygen and hydrogen, wonder of wonders, we have water, a crucial ingredient for life's recipe. Next add a dash of carbon and a touch of sulfur. Finally, bolster the formula with some nitrogen and phosphorous.

That's only the beginning of an inorganic base essential for intelligent life! And don't forget, DNA's pre-programmed information, is an imperative for cell function and reproduction!

Just as life, as we know it, does not generate spontaneously, it has yet to be observed in outer space although odds suggest some form of intelligent life likely exists somewhere else in the universe.

So what's the probability of finding a free-floating space station offering an environment capable of generating organic life by chance?

This unlikely event demands a confluence of cosmic coincidences!

For starters, a life-friendly ecosystem requires factors converging by chance in a split second of time, at a location no more than a micro-mini speck in space, complete with a convenient and readily available array of those life-essential elements.

This series of conditions friendly to life, all within the cross hairs of one small dot in an infinity of space, beckons recognition of something beyond the luck of the draw. Logic teams with science and Scripture, pointing to a Master Designer, somebody *"bigger than you and I."*

"It seems as though somebody has fine-tuned nature's numbers to make the Universe ...The Impression of design is overwhelming...We are truly meant to be here." [20]

The odds of an explosion, creating an environment friendly to the production of organic life, is less likely than all earth's citizens solving a Rubik cube puzzle, simultaneously, in less than a minute---then repeating the exercise, without error, a million more consecutive times.

For starters, earth's life-friendly ecosystem thrives on balanced land/water ratios, a reliable supply of fresh water, all nestled within an atmospheric envelope with delicately matched proportions of oxygen, carbon dioxide, ozone, and nitrogen.

The mass, color, location, and luminosity of stars; earth's orbit inclination and axis tilt; surface crust thickness; gravity that keeps feet planted securely; and magnetic fields combine to suggest *"if masses did not attract each other, there would be no planets or stars, and once again it seems that life would be impossible."* [21]

Explorers from the Blue Planet home base, survive in space's hostile environment by carrying their own life-support systems. Fragile human life hinges on access to oxygen, water and food.

Air deprivation guarantees suffocation in short minutes.

Without water, a person can last several days. Too much water drowns its victims. Without water, death by dehydration awaits.

Starvation takes longer but is just as certain. Take away nutritious food, and the strongest person might live but a few, short weeks.

Temperatures too hot or too cold accelerate the death process. Too much, too little, too far, too near, too late, too soon—any factor out-of-kilter and life on earth could not exist. Move the sun much closer, and the earth would be scorched; much farther and life would fade into deep freeze.

And don't forget that ozone mantle, wrapped protectively around earth, shielding life from overpowering ultraviolet radiation.

Consistent doses of sunlight, radiating beams of ultraviolet and infrared, sustain life. If the sun's relationship to the electromagnetic spectrum shifted imperceptibly, the chance of life could vanish.

Instead of blistering heat or deadly radiation, energy from the sun comes calibrated in a range maximizing a life-friendly environment. Sunlight radiates a collateral bonus in the riot of colors that embellish environments. Restful sky-blues, backlighting forests of multi-hued greens, define the landscape.

A rainbow of kaleidoscopic accents ranging from pastel shades of shimmering pinks and lavenders to crimson-golds, trace the arc of a daily rising and setting sun. Shifting combinations of the sun's rays conspire to induce psychological peace.

Despite somber predictions that the raging inferno at the heart of the Solar System is certain to destroy itself someday in a blazing conflagration, the sun ignores dire predictions and keeps *"rollin' round heaven all day."*

"Almost everything about the basic structure of the universe...the fundamental laws...of physics and the initial distribution of matter and energy...is balanced on a razor's edge for life to occur." [22]

At the very least, life generating spontaneously from non-life, would demand a reducing atmosphere, where atoms and molecules bond with hydrogen rather than oxygen.

The amino acids essential to evolution's formula for generating organic life requires an oxygen-free, reducing environment—a condition missing when spontaneous generation was supposedly doing its magic in Darwin's *"warm little pond."*

Oxidation renders spontaneous generation of life impossible. Yet, evidence points to the fact that *"Oxygen was likely present in the early earth's atmosphere."* [23]

Spontaneous generation postulates earth's atmosphere lacked oxygen when life formed. Quite the contrary, there is *"…strong evidence that oxygen was present on the earth from the earliest ages…Significant levels of oxygen would have been necessary to produce ozone which would shield the earth from levels of ultraviolet radiation lethal to biological life."* [24]

A fully oxidizing nature, characteristic of earth's prehistoric atmosphere, is hostile to the assembly of new life, creating a chemical dilemma. Atoms and molecules tend to bond with oxygen atoms while free oxygen inclines to oxidize organic compounds, destroying the chemical building blocks of life.

If the *"early atmosphere was oxygen-free…then there would have been no protective ozone layer. Any DNA and RNA bonds would be destroyed by UV radiation… Either way, oxygen is a major problem."* [25]

"…All experiments simulating the atmosphere of the early earth have eliminated molecular oxygen…Oxygen acts as a poison preventing the chemical reactions that produce organic compounds…If any chemical compounds did form, they would be quickly destroyed by oxygen reacting with them…" [26]

"Even if oxygen was not present in the early earth's atmosphere, the absence of oxygen would present obstacles to the formation of life. Oxygen is required for the ozone layer which protects the surface of the earth from deadly ultraviolet radiation. Without oxygen this radiation would break down organic compounds as soon as they formed." [27]

Microbiologist Michael Denton reasons *"In an oxygen-free scenario, the ultraviolet flux reaching the earth's surface might be more than sufficient to break down organic compounds as quickly as they were produced…In the presence of oxygen, any organic compounds formed on the early Earth would be rapidly oxidized and degraded…*

"The level of ultraviolet radiation penetrating a primeval oxygen-free atmosphere would quite likely have been lethal to any proto-organism possessing a genetic apparatus remotely resembling that of modern organisms. What we have then is a 'Catch 22' situation.

"If we have oxygen, we have no organic compounds but if we don't have oxygen we have none either." [28]

Ultimate bottom line: Life evolving by lucky accident faces a classic lose/lose scenario---life could never evolve in an atmosphere with oxygen; but once formed, life could not survive without oxygen!!!

Life's origin and survival also demands water---a rare commodity in the Solar System!

Its axiomatic: no water, no life!

A nagging problem lingers: how could an explosive big bang in space wrap earth in a wet blanket but leave the circling moon and planets as high and dry as the desert sands? Conventional science does not expect to find life flourishing on a bone-dray planet.

The creation story describes an originally empty and formless earth with darkness *"over the surface of the deep and the Spirit of God was hovering over the waters."* [29]

Viewed from space, our Blue Planet reflects a bright blue hue. No wonder---supposedly 70% of the earth is covered with water. So much water, if the earth's land crust was flattened, water hundreds of feet deep would cover its surface. Scholars have drawn startling conclusions about this natural water supply.

For example, 96.5% of earth's water consists of ocean marine water and 0.97% brackish water (An incidental aside is the report suggesting 95% of the total fossil record consists of marine life).

Thanks to NASA and astronauts
like Colonel Douglas H. Wheeler, earthlings are treated
to rare glimpses of earth from the heavens.

This is how Egypt, the Red Sea, Sinai Peninsula,
Jordan River and the Sea of Galilee appear from space.

That leaves 2.53% of our entire water supply as fresh water. This comparatively minuscule amount is allocated as: 69.6% glaciers and permanent snow; 30.1% ground water; 0.29% lakes, marshes, and swamps; 0.05% soil moisture; 0.04% atmosphere; 0.006% rivers; and 0.003% living organisms. [30]

Whether evolutionist belief in a series of lucky accidents or creationist belief in the eternal God, faith dominates origin of life scenarios.

"Either life was created on the earth by the will of a being outside the grasp of scientific understanding, or it evolved on the planet spontaneously, through chemical reactions occurring in nonliving matter lying on the surface of the earth.

"The first theory…is a statement of faith in the power of a Supreme Being…The second theory is also an act of faith…assuming that the scientific view of the origin of life is correct, without having concrete evidence to support the belief." [31]

Rejecting belief in an all-powerful Creator as faith-based "religion" while characterizing assumptive, unproven, faith-based secular theory, as "science" epitomizes intellectual hypocrisy.

Regardless of the label, marketing a poison pill as a vitamin does nothing to enhance public health!

"The chances that life just occurred are about as unlikely as a typhoon blowing through a junkyard and constructing a Boeing 747." [32]

Life does not create itself from non-life---not millions of years in the past! Not today! Not ever!

Matter and energy don't result from "nothing" exploding! When a computer crashes, intelligent information never uploads itself!

Evolution fallacy lacks foundation, function or future. Its consistent only in its incoherent irrationality.

With no place else to go, life resulting from a destructive explosion then an accident in some warm pond, retreats to mystical shadows of deep time, impervious to the *"flaws"* and *"holes'* in Darwinian rhetoric.

IV

Superstitious Nonsense

Spontaneous Generation?

"To get a cell by chance would require at least one hundred functional proteins to appear simultaneously in one place. That is one hundred simultaneous events each of an independent probability which could hardly be more than 10^{-20} giving a maximum combined probability of 10^{-2000}." [1]

Michael Denton

© Nagy-Bagoly Arpad

Charles Darwin claimed, *"The chief distinction in the intellectual powers of the two sexes is shewn by man attaining to a higher eminence in whatever he takes up, than woman can attain—whether requiring deep thought, reason, or imagination, or merely the use of the senses and hands."* [2]

Charles Robert Darwin lacked the faintest clue or the least scintilla of scientific evidence explaining the origin of first life on Planet Earth!

More than 150 years after he went public with his dubious speculations, the what, when, where, and how the simplest living cell managed to create itself accidentally from non-living, pre-existing matter continues to baffle!

Even blessed with research technology unavailable top Darwin, the gap in flawed logic continues to plague evolution theory as confronted with nagging evidence shortfall.

© Andrly Nekrasov

Unaware of DNA or even a cell's nucleus, Darwin imagined life created itself, by accident, in some *"warm little pond."* [4]

What would there be to evolve unless that first life managed to create itself by the magic of spontaneous generation---pre-loaded with DNA information from some unknown source?

Darwin had never heard of DNA, or even a cell's nucleus, so he waved his wand imagining that the magic of self-creation took place by accident in the some ancient, unidentified, *"warm little pond."*

And evolution's theoretical beginning is not miracle based? The "warm little pond" scenario seems even less likely than finding an iceberg floating in a desert mirage!

"A number this large [see Carl Sagan's estimate, supra] *is so infinitely beyond one followed by 50 zeroes (Borel's upper limit for such an event to occur)…There is, then according to Borel's law of probability, absolutely no chance that life could have 'evolved spontaneously' on the Earth."* [3]

Spontaneous generation of a living cell, by chance, from inorganic matter, overwhelms imaginations. Just how could a cell, without evidence of prior ancestry, arrive from nowhere, pre-loaded with the genetic information carried in a single gene?

© Vlad61

"All the higher mammals are probably derived from an ancient marsupial, and this through a long line of diversified forms, either from some reptile-like or some amphibian-like creature, and this again from some fish-like animal." [5]

Richard Hutton, Executive Producer of the controversial PBS TV series, *"Evolution,"* was asked, *"What are some of the larger questions still unanswered by evolutionary theory?"*

He replied: *"The origin of life. There is no consensus at all here---lots of theories, little science. That's one of the reasons we didn't cover it in the series. The evidence wasn't very good."* [6] Maybe "worthless" rather than *"not very good?"*

"It is often said that all the conditions for the first production of a living organism are now present, which could ever have been present.

"But if (and oh! what a big if!) we could conceive in some warm little pond, with all sorts of ammonia and phosphoric salts, light, heat, electricity, &c., present, that a proteine compound was chemically formed ready to undergo still more complex changes, at the present day would be instantly devoured or absorbed, which would not have been the case before living creatures were formed." [8]

Lacking the first clue as to the what, how, when, where and why, of first life's origin, Darwin conceded, *"Science as yet throws no light on the far higher problem of the essence or origin of life."* [7]

The best Darwin could do to explain this *"far higher problem"* was to side-step the non-explanation's dubious adequacy. Instead, evolution's guru chose to rock the 19th century world by releasing the first edition of his *Origin of Species* in 1859.

Claiming all life began with that mysterious cell from nowhere, he imagined the "miracle" was followed by a fanciful series of incremental changes over mega chunks of deep time.

If Darwin is taken at his word, the resulting do-it-yourself "biological transit stops" supposedly produced every species of plants and animals known to man, eventually replacing all ancestor species with these radically new and different body plans.

School kids explain, matter-of-factly, that two atoms of hydrogen joined to a single atom of oxygen build a water molecule---it always has, and always will. Yet, minds swearing allegiance to the spontaneous generation fallacy are unable to conjure up the recipe mix for the quite hypothetical and always elusive "prebiotic soup."

Despite the unresolved mystery dooming the origin of first life by chance rather than by intelligent design, the handiwork of a Designer, Darwin allied his faith with philosophers like Aristotle, who preached spontaneous generation as some pagan brand of superstitious gospel.

Then along came Frenchman Louis Pasteur with his scientific demonstrations that shattered the "spontaneous generation" myth.

Today, the formula for first life continues to baffle devout evolutionists who can't begin to replicate the simplest cell and are at a loss to explain the source of genetic information packed into that first cell's DNA.

Lacking elsewhere to turn, evolution's more dedicated shills tout outer space as a possible ultimate source of life on earth. Even if plausible, a pair of problems remains: how did a cosmic life form survive the hazards of space travel and how did it manage its own life's accidental kick-start?

Evolutionism's recipe for Formula One evades hot pursuit.

By default, it is left to Mother Nature's whims and Pappy Time's antiquity to parent first life.

As to evidence of the chance appearance of that the first ever, living cell from non-living matter, evolution theory draws a blank screen!

With even Darwin questioning *"chance"* as a serious factor in *"the world as we know it,"* why not confront the ultimate challenge to human intelligence and create life in a lab from non-living matter?

If life from unintelligent non-life could result theoretically from an accidental whim of nature, then why couldn't human intelligence be recruited to design and create a simple, living cell from inert matter, duplicating the secret of first life's launching pad allegedly hiding in Darwin's *"warm little pond"?*

The challenge beckoned audacious minds!

Without the magnifying power of electronic microscopes, 19th century scientists dismissed a living cell as nothing more than a protoplasm blob.

Then along came Gregor Mendel and his garden of colorful blossoms, introducing science to the world of genetics. And that world soon staggered conventional thought!

That presumed simple "blob" opened eyes to unimagined complexity.

What appeared through the electron microscope's lens was a throbbing piece of molecular machinery, complete with a nucleus packed with genetic information---a variety of proteins, all wrapped in a membrane and surrounded by a cell wall.

"The tiniest bacteria cells are incredibly small, weighing less than 10^{-12} gms, each is in effect a veritable microminiaturized factory containing thousands of exquisitely designed pieces of intricate molecular machinery, made up altogether of one hundred thousand million atoms, far more complicated than any machine built in the non-living world...nor is there the slightest empirical hint of an evolutionary sequence among all the incredibly diverse cells on earth...

"The complexity of the simplest known type of cell is so great that it is impossible to accept that such an object could have been thrown together suddenly by some kind of freakish, vastly improbable event. Such an occurrence would be indistinguishable from a miracle."[10]

The single-cell *Mycoplasma,* a microorganism without a cell wall, the *"simplest known self-reproducing life form,"* carries 482, life-directing genes. [11]

The blind chance of spontaneous generation producing the complete formula of molecules, amino acids, and proteins essential for a cell only one-tenth the size of *Mycoplasm hominis H. 39,* is less than one in $10^{340,000,000}$.

That's 10 with 340 million zeros after it. [12]

So the stage was set for a scientific challenge pitting nature's "chance" versus intelligent "design." Could something similar to *Mycoplasma* be reproduced in a laboratory under the auspices of human minds?

Stanley Miller and Harold Urey stepped to the plate in 1953, intent on answering the challenge---the same year the DNA double helix string of information housed in living cells grabbed international headlines!

This innovative duo shaped their experiment following a theoretical trail blazed by Russia's Alexander I. Oparin's and Britain's J.H.S. Haldane's attempt to rescue spontaneous generation ideology from history's dust bin of dead-end, trashed ideas.

Adding intelligence to the formula struck at the heart of an equation devised to authenticate impossible accident! Even if successful, the intervention of human thought sabotaged the credibility of a process intended to confirm the non-intelligent origin of life!

Amino acids, the building blocks linking cell proteins, have been synthesized in laboratory environments. But the creation of a full complement of proteins essential to life, from laboratory-built amino acids, is as futile as breaking the sonic barrier riding a broomstick.

There is more: amino acids can't form in the presence of oxygen!

Sidestepping this hurdle, the Oparin-Haldane hypothesis proposed earth's original atmosphere consisted of carbon dioxide, carbon monoxide, ammonia, methane, hydrogen and water vapor---but without oxygen.

Consequently, a reducing atmosphere dictated laboratory methodologies used to attempt the creation of a living cell from non-living matter!

Oparin-Haldane theory supposed inorganic *"…chemicals combined to form organic compounds, such as amino acids, which in turn combined to form large, complex molecules, such as proteins, which aggregated to form an interconnecting network and a cell wall."* [13]

Applying this formula as a basis for manufacturing life in a laboratory, the Miller/Urey team attempted to create a reducing atmosphere by circulating a high energy spark through methane, ammonia, and hydrogen gases and a circulating hot water vapor.

The process produced *"a small mass of black tar"* along with *"a condensed red liquid"* containing some amino acids. The nagging problem persists: the experiment *"…only works as long as oxygen is absent and certain critical ratios of hydrogen and carbon dioxide are maintained…"* [14]

Subsequent experiments using ultraviolet radiation produced *"nineteen of the twenty biological amino acids and five nucleic acid bases of DNA and RNA."* [11]

The Miller/Urey duo required a torturous trail of happenstance with each step unaccountably fostered in the prebiotic soup.

They counted on energy from lightning, earthquakes, volcanoes, and the sun's rays to trigger chemical reactions with atmospheric gases such as methane, ammonia, hydrogen, ethane and water vapors, conveniently converting them to amino acids, fatty acids and sugars in ancient oceans.

Relying on the luck of the draw, these compounds could then theoretically link up to form larger protein and DNA molecules, ultimately becoming *"the first true cell"* capable of *"metabolism, genetic coding, and the ability to reproduce"* when wrapped with a membrane. [15]

One public school text, buying into the spontaneous generation line, misled students with a string of imaginative speculations.

"…Primitive earth may have had an atmosphere largely of hydrogen which was later lost to space. A secondary atmosphere may have included ammonia, methane, water, and hydrogen sulfide…Ultraviolet light from the sun, electrical storms, and decay of radioactive elements may have provided the energy to combine these molecules as sugars and amino acids.

"Amino acids could have combined to form proteins…" [16]

"May have" and *"could have"* don't disguise gross speculation!

Could lightning, heat from volcanoes and the sun's ultraviolet rays have actually *"…affected gases in the primitive earth's atmosphere and changed them into more complicated organic compounds…"* such as fatty acids, amino acids, sugars, and nucleotides?

Then in turn, by magical hocus pocus, *"accumulated in the ocean and then linked up with each other to form very complex molecules…"* such as lipids, peptides, carbohydrates, polynucleotides and eventually combined to form *"amazingly complex proteins?"* [17]

"Urey and Miller assumed that methane was plentiful in the early earth's conditions. If this is true, the sun's ultraviolet light would have caused hydrocarbons to form and absorb in the clay at the bottom of the ocean. The deposits from Precambrian periods should then contain significant hydrocarbons or remains of carbons, as well as some nitrogen containing compounds. None of these are present in these deposits." [18]

Contrary to assumption, evidence confirming plentiful methane in the early earth environment continues elusive.

"…Oxygen is necessary to protect proteins and DNA from the sun…Living organisms [bacteria] and organic molecules [amino acids, proteins and DNA] need the protection from ultraviolet radiation provided by an ozone screen [which is derived from oxygen]. [19]

"…If even trace amounts of molecular oxygen were present, organic molecules could not be formed at all." [20]

"Since living organisms [bacteria] *and organic molecules* [amino acids, proteins] *need the protection from ultraviolet radiation provided by an ozone layer* [which is derived from oxygen] *yet the presence of oxygen,* [in the atmosphere] *prevents the development of such living systems and biological molecules* [amino acids], *this would constitute a catch-22 in the model."* [21]

Oxygen is the critical component of today's atmosphere!

Oxides in the rocks suggest oxygen was present in ancient atmospheres. *"Iron oxide minerals have been found in Greenland, dating to 3.9 billions years ago. The presence of oxides suggests that oxygen was present at the time."* [22]

Overeager celebrants initially interpreted the experiment's result as a break-through in creating virtual life from non-life in a test tube.

Realists recognized much less. Brilliant human minds hit the wall---over their heads in a realm reserved for a *"Superior Rationality."*

The Miller-Urey experiment faced *"…withering criticism from chemists for ignoring the role of competing and destructive cross-reactions with chemical ions that would be expected in any hypothetical ocean or pond. These reactions would have tied up or terminated any growing polymer-chain."* [23]

The brilliant Miller/Urey team did not create life. Nor did they prove life could have originated spontaneously from inorganic matter. Innovative human genius couldn't do the trick!

Finite minds, capable of creating computers, have never successfully designed and built a living cell from non-living, inorganic matter, much less shaped a strand of information-packed DNA.

The 1953 lab experiment fizzled.

Early in 2011, a team of Japanese scientists announced a five-year plan to access cloning technology in an effort to introduce the extinct mammoth to the twentieth century. The plan calls for replacing the nuclei from an elephant's egg cell with DNA extracted from mammoth tissue preserved in a Russian lab, and then placing the redesigned egg in an elephant's uterus.

Whether or not successful, cloning, like gene splicing, is not the creation of new life from non-living matter but is the result of human intelligence designing life forms by mixing precisely selected DNA with pre-existing living tissues.

Redundant propaganda touting unproven assumptions as scientific *"fact"* camouflage congenital defects plaguing evolution dogma.

Assumptions provide a shifting sand foundation for a theory every bit as assumptive as the November 24, 1859 day when *The Origin of Species* hit London streets.

Assertions built on assumptions don't equate fact.

Superstitious nonsense blossoms when the assumption virus invades reasoning. Clinging to assumptions as fact is a faith exercise!

Science, like religion, can be vulnerable to compromise by assumption.

Media in a free society champions free speech in its pursuit of truth. Still, subtle enticements, beckoning from evolution's *"working hypothesis,"* can turn "fact" upside down. *USA Today* fell prey to the trap in its August 9, 2005 edition, by exalting evolution and taking a swipe at "Intelligent Design," a troubling nemesis of the doctrine.

"It is the cornerstone of modern biology. Though there are various 'missing links' in the evolutionary chain, it has never been refuted on a scientific basis." [24]

Is something out of kilter here?

Its strange irony that a team of intelligent minds composed phrases, designed a layout, printed and distributed thousands of copies while insisting human brains self-created, by random accident, from some undetermined, unintelligent source---without design or designer.

What precisely did Darwin *"postulate"* that *"scientists have confirmed?"* Brushing ponderous phrases aside, if Darwin's postulates were a publicly traded stock, its pending market collapse should be imminent.

British scientist Gerald A. Kerkut, has acknowledged evolution is riddled with unproven assumptions that *"…by their nature are not capable of experimental verification."*

The list of seven awarded top billing to the assumption *"that non-living things gave rise to living material, i.e. spontaneous generation occurred."* [25]

Kerkut's six other eye-popping assumptions do nothing to fortify evolution's credibility as *"confirmed"* science.

"The second assumption is that spontaneous generation occurred only once…The third assumption is that viruses, bacteria, plants and animals are all interrelated. The fourth assumption is that the Protozoa gave rise to the Metazoa.

"The fifth assumption is that the various invertebrate phyla are interrelated. The sixth assumption is that the invertebrates gave rise to the vertebrates. The seventh assumption is that within the vertebrates the fish gave rise to the amphibia, the amphibia to the reptiles, and the reptiles to the birds and mammals…" [25]

Kerkut dismantles evolution's grand scheme of some organic chain's gradual evolutionary process. His long unresolved roster of assumptions, lurk menacingly, nagging persistently at the fringes of Darwinian thought.

Kerkut's candid assessment concludes, *"The 'General Theory of Evolution' and the evidence that supports it is not sufficiently strong to allow us to consider it as anything more than a working hypothesis."* [26]

Evolution's legacy rests on this pile of shifting, assumptive sand!

Lively imaginations grease the wheels of assumption's dreams. But imagination is an inadequate substitute for evidence. Assumptions become presumptions providing transparent cosmetic cover for conjecture.

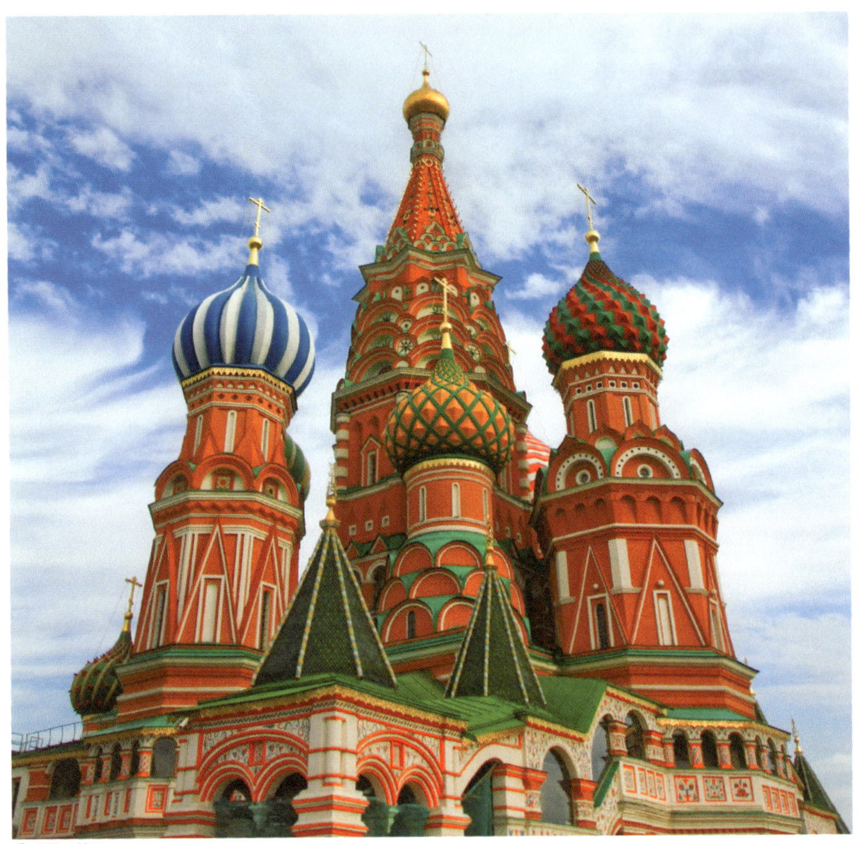

**Colorful architecture results from intelligent design
not accidental, spontaneous generation.**

Until verified by evidence, assumptions melt like wax caressed by red rays of the noon-day sun.

Nothing of significance has changed since Kerkut's reality checks.

Speculation that some primordial soup mixed with doses of heat, cold, water, and lightening somehow delivered genetic information creating a living cell proves nothing other than vivid, unrealistic imaginings.

Post-1859 breakthroughs have yet to rescue Darwin's ideas from its congenital *"flaws"* and *"holes"* in order to warrant christening evolution a *"fundamental fact of biology."*

British evolutionist, Sir Fred Hoyle concluded, *"the notion of life---with its incredibly intricate genetic code---originating by chance in some sort of primordial organic soup, was…'nonsense of a high order.'"* [27]

A class-five hurricane scrambles and devours telephone lines marking its fickle path of raging fury. No one has seen this fearsome force shape the jumble of metal scraps, strewn helter-skelter in its twisted wake, to accidentally design and create a fully-functioning personal computer—much less manufacture self-replicating models!

Accidental creation of a computer by a hurricane's havoc seems less than farfetched compared to the chance of a living cell spawning itself spontaneously from warm water!

Nobel laureate Sir Francis Crick observed:

"An honest man, armed with all the knowledge available to us now, could only state that in some sense, the origin of life appears at the moment to be almost a miracle, so many are the conditions which would have had to have been satisfied to get it going." [28]

Louis Pasteur (1822-1895), the French chemist who fathered microbiology and introduced vaccines to combat anthrax and rabies, discarded medieval superstition, demonstrating the vacuity of the belief that inorganic matter spontaneously generates life.

His experiments staked out findings that pruned out the heart of naturalism's taproot. Confirming that life begets life and like begets like, he demonstrated:

"…The impossibility of the appearance of life from non-living matter…To bring about spontaneous generation would be to create a germ. It would be creating life…God as author of life would then no longer be needed. Matter would replace Him." [29]

Hardly a year had gone by after publication of the first edition of *Origin of Species,* before Darwin admitted to troubled, second thoughts.

Writing to his American friend and Harvard botanist, Asa Gray, the aging Darwin dropped a verbal bombshell.

After proposing a theory built on *"chance,"* evolution's big gun, still stubbornly refusing to acknowledge *"Design"* in nature, confessed to finding his personal thinking in a *"hopeless muddle."*

Darwin's pause for second thoughts was justified!

Something was strangely out-of-kilter with the view that a living cell was a "simple" blob of protoplasm, and capable of creating itself, by chance, from non-living matter.

More than a century later, molecular biologist Michael Denton set the record straight, dismissing chance in the cell formation process!

Denton characterizes spontaneous generation of a living cell as both *"freakish"* and *"vastly improbable."*

More than the simplistic blob of protoplasm perceived in Darwin's day, every living cell contains unique DNA information that manages living systems. Denton reasoned it would be *"indistinguishable from a miracle"* if a cell had been *"thrown together suddenly by some kind of freakish…event."* [31]

"It is the sheer universality of perfection, the fact that everywhere we look, to whatever depth we look, we find an elegance and ingenuity of an absolutely transcending quality, which so mitigates against the idea of chance." [32]

Aware of the primitive nature of scientific technology in the mid-nineteenth century, Denton cut Darwin some slack.

"There is little doubt that if this molecular evidence had been available one century ago it would have been seized upon with devastating effect by the opponents of evolution theory…and the idea of organic evolution might never have been accepted." [33]

Mathematically, the odds against life by "chance" equate impossible!

Regardless, the theory's media patrons tend to flak, hype, and shill evolution's obsolete myths in a blizzard of clichés, confusing the public by awarding sham "science" an undeserved place of honor in the pantheon of respectability.

Mark Twain's tongue-in-cheek humor resonates!

"There is something fascinating about science. One gets such wholesale returns of conjecture out of such a trifling investment of fact." [34]

<h1 style="text-align:center">V</h1>

<h2 style="text-align:center">Paging Sherlock Holmes</h2>

Missing Links

"To take a line of fossils and claim that they represent a lineage
is not scientific hypothesis that can be tested,
but an assertion that carries the same validity as a bedtime story—
amusing, perhaps even instructive, but not scientific." [1]
Henry Gee

© Galyna Andrushko

**The colorful chasms of Arizona's Grand Canyon
reveal artifacts of deep time but little
evidence of missing links.**

Charles Darwin recognized his theory teetered in serious jeopardy unless confirmed by an *"inconceivably great"* quantity of *"transitional links"* yet to be discovered in the fossil record.

"If my theory be true, numberless intermediate varieties, linking closely together all the species of the same group, must assuredly have existed" [2]

"The number of intermediate and transitional links between all living and extinct species must have been inconceivably great." [3]

Several million fossil finds later, the cupboard of *"transitional links"* continues virtually bare. Using Darwin's own rules---game over!

End of story?

Diehards haven't given up---goal posts keep moving.

While bypassing the how, when, and where, Charles Darwin spun a tale of a *"finely graduated organic chain"* [4] of life evolving gradually from that still unexplained first living spark to a plethora of diversified complexity in a grandly conjectured "tree of life."

Choice words highlighted unproven assertion.

Not yet apparent in living systems, evolution's last best hope for evidence such an *"organic chain"* is anything more than wishful thinking, Darwin reasoned proof must be hiding still in the dry bones of past lives.

Optimistic predictions have not fared well exposed to the shovels and microscopes investigating post-Cambrian evidence. In a game of paleontological hide-and-seek, undiscovered fossil fragments of lives that never existed have yet to be found!

Intensive quests have turned up little but frustration's heat.

*****

True to his word, Darwin accessed his fertile *"imagination,"* conjuring up visions of never-seen, fictional critters intended to illustrate his radical idea. Not to be outdone by English storyteller, Charles Dickens, Darwin ventured to suggest an attention-getting whopper!

"Some extremely remote progenitor of the whole vertebrate kingdom appears to have been hermaphrodite or androgynous." [5]

Ultimately, the organic chain of quaint imaginings gradually drifted past the *"vertebrate…hermaphrodite"* phase leading quite accidentally to a dubious depiction of human ancestors burdened with a *"tail"* and a head crowned with *"pointed ears."*

"Man is descended from a hairy quadruped, furnished with a tail and pointed ears, probably arboreal in its habits…" [6]

"Early progenitors of man were no doubt once covered with hair, both sexes having beards; their ears were pointed and capable of movement; and their bodies were provided with a tail…" [7]

This is not make-believe movie madness but Darwin's very serious overreach to demonstrate the accidental potential of his idea if given enough time---all descending from that original cell that theoretically débuted on earth at a time and place he knew not how, when or where.

So what was the legacy of Darwin's *"imagination?"*

His bold conjecture called for the constant shifting of unstable life forms, moving in a series of blind transititions to unpredictable *"biologic transit stops."*

The English naturalist postulated intermediate organic life forms climbing up an imagined taxonomic tree.

He envisioned all living plants and animals evolving up a taxonomic tree to ever more complex organisms from that original "simple" cell, compliments of *"natural selection."* The process envisioned millions of miniscule increments of genetic change over mega chunks of deep time.

59

© Chuck Nelson

Supposedly, that first life progressed gradually up the taxonomic ladder through a yet-to-be-proved evolutionary sequence: *"Fish-like," "Aquatic animal," "Amphibian-like creature," "Reptile-like," "Higher mammals," "Old World division of the Simiadae;"* with *"Man"* emerging at the top of the heap.

Evidence destroys this assumption! Genetic adjustments can be real but the changes work downstream, down the taxonomic ladder, not up.

Contrary to evolution theory that takes off from the trunk of a single *"tree of life,"* the fossil record exposes multi- thousands of distinctly different plant and animal life forms that appear, abruptly without evidence of prior ancestry.

Reliable evidence of an organic chain of ancestral life forms, transiting one kind-to-a new and different kind, doesn't exist!

At least 7,640 fully-formed, mostly marine animal species, appeared worldwide, in a simultaneous *"sudden leap,"* saturating the Cambrian. [8]

The *"almost abrupt appearance of the major animal groups,"* representing as many as 50 phyla during the Cambrian Period, ranks as *"…one of the most difficult problems in evolutionary paleontology…"* [9]

60

Cambrian fossils demolish Darwin's core conjecture!

© Jacek Jasinski

© Robert Paul Van Beets

**Remnants of fossilized trees from an ancient forest,
swept away in a flood, decorate Arizona's desert landscape.**

No evidence suggests these trees were transitional.

Unrelated phyla appear abruptly, across-the-board, rather than evolving gradually, starting with a single cell. [10]

Multi-celled animal embryos, no bigger than a grain of sand, have been discovered in China, preserved in calcium phosphate and dated at the edge of the Cambrian/Precambrian time frame—marking the prolific explosion of organic life during which *"virtually all the major animal body plans seen on Earth today blossomed in a sudden riotous evolutionary springtime."* [11]

The Cambrian explosion of multi-celled organisms covers the waterfront as to invertebrate animal phyla, both living and extinct. Arthropods, mollusks, and echinoderms—there they are, without a fossil clue as to prior *"numerous, successive, slight modifications."*

"Most orders, classes, and phyla appear abruptly, and commonly have already acquired all the characters that distinguish them." [12]

For all of the animal phyla to appear in one single, short burst of diversification is not an obviously predicable outcome of evolution." [13]

The presence of fully-formed fossils, rather than intermediates, conventionally dated more than 500 million years before the present, seriously impairs evolution's expectations.

"With few exceptions, radically new kinds of organisms appear for the first time in the fossil record already fully evolved, with most of their characteristic features present." [14]

Henry Gee discounts evolution's supposed organic chain of ancestry as *"…a completely human invention created after the fact, shaped to accord with human prejudices…Each fossil represents an isolated point, with no knowable connection to any other given fossil, and all float around in an overwhelming sea of gaps."* [15]

Far from supporting descent from a chain of organic intermediates, the Cambrian and Pre-Cambrian fossil record offers a virtual "paleontological desert," barren of readily identifiable fossil ancestors.

All irreducibly complex animal and plant life forms descending from some mysterious, original single-cell first life as conjectured by evolution theory doesn't match Cambrian fossil reality.

Choice rhetoric highlights unproven assumption. The naturalist
conjectured that chunks of deep time made anything possible---even
mathematically impossible leaps over vast chasms of biological diversity.

Cambrian fossils deliver a triple whammy to the theory: fully-formed
life existed from the get-go without evidence of prior fossil ancestry;
intermediates elude discovery; and many descendant species carry
remarkable resemblance to their theoretically extinct ancestors.

Late in the nineteenth century, paleontologists Edward Drinker Cope
and Othniel Charles Marsh squandered personal reputations competing for
fame and fortune in a "Bone Wars" vendetta!

Their efforts harvested tons of dinosaur remains, strewn in fossil
cemeteries across the high prairies of Colorado and Wyoming. Bitter rivalry
raged between the two adversaries from 1877 to 1892.

No question, multi-ton finds of gigantic dinosaur bones confirmed
extinction of a once monstrous species. But the petty dispute between two
adversaries exhausted professional careers and personal fortunes.

Several-mile stretches of massive fossil cemeteries indicated where lush
forests once stood.

Marine fossils in the mix at elevations as high as 7,881 feet in the general
area of Morrison, Colorado left evidence of what must have been rapid
burial in Dakota sandstone. Sudden hydraulic action, without precedent in
the modern world, prevented decay.

Cope and Marsh were not alone in straddling dinosaur backs in hopes of
professional recognition and financial dividend.

Paleontologist Hermann von Meyer led the pack, ready to cash in with
the remains of *Archaeopteryx*, a pigeon-sized imprint of bones unearthed in
Southern Germany's Solnhoffen Quarry.

© Stelian Ion

Did an octopus descend from some still missing link?

Earlier, French paleontologist Geoffrey Saint-Hilaire, floated the idea that birds possibly evolved from reptiles.

The concept ignored radical dissimilarities between the two animal types (eg. warm-blooded birds v. cold-blooded reptiles). Darwin loyalists, eager to bridge the fossil record's yawning gaps where intermediates failed to show, seized upon the *Archaeopteryx* remains, etched in Solnoffen limestone slabs, as evidence of reptile-to-bird transition.

Doubters, then and now, question the connection!

Andreas Wagner, discoverer of a dinosaur fossil he named *Compsognathus*, dismissed the claim, warning:

"Darwin and his adherents will probably employ the new discovery as an exceedingly welcome occurrence for the justification of their strange views upon the transformation of the animals. But they will be wrong." [16]

Alan Feduccia scoffed at the dinosaur-to-bird scenario.

"The theory that birds are the equivalent of living dinosaurs and that dinosaurs were feathered is so full of holes that the creationists have jumped all over it, using the evolutionary nonsense of 'dinosaurian science' as evidence against the theory of evolution...To say dinosaurs were the ancestors of the modem birds we see flying around outside today because we would like them to be is a big mistake." [17]

So was *Archaeopteryx* just another extinct bird species or even a pigeon-sized dinosaur?

Since the 10-inch wingspan *Pterosaur* was also discovered nearby in Sondheim, Bavarian limestone, could *Archaeopteryx* have been a relative?

No sooner had the twenty-first century dawned than the bird-to-dinosaur scenario took a hit from another counterfeit fossil.

This time, the victim was much more than a bit player in the drama.

Photos of fossil remnants trumpeted as *"...a missing link between terrestrial dinosaurs and birds that could actually fly...a true missing link in the complex chain that connects dinosaurs to birds"* adorned full-color pages of the November, 1999 issue of the prestigious *National Geographic.* [18]

Big bucks backed the journal's investment in "authentic" antiquity! The endorsement came loaded with monumental consequences given the journal's multi-million subscriber circulation and its proud tradition of editorial excellence!

But despite the pretty pictures and the exuberant tone implying authenticity, by January of the new millennium, the speculative dreams of the sponsor came crashing to the earth. The costly find, imported from China's Liaoning Province, proved to be nothing more than a carefully concocted swindle.

In a slight-of-hand game of "Pin the Tail," some enterprising huckster hitched the tail of a long-deceased dinosaur onto the fossil remains of an equally unsuspecting bird.

The only thing the weird dino/bird fossil combo achieved was the egg it smeared on the faces of over-eager evolution enthusiasts.

Lemmings march mindlessly off cliffs in a mysterious dance of death. Remarkably, the marvelously more sophisticated human mind is vulnerable to multiple brands of cultural coercion.

A French game show instructed contestants to jolt fellow contestants with up to 450 volts of electricity when they heard wrong answers. Despite the anguished cries of the "victims," four of five participants reacted to noisy shouts from the audience, pushing them to inflict unconscionable pain on other human beings.

Unaware the event was a sham, without any surging voltage, and that writhing "victims" were actors selected for a scientific test, all but 16 of the 80 contestants followed instructions mechanically, irrespective of the perceived "hurt" they inflicted following instructions.

Not exactly a vote-of-confidence as to freethinking independence and gracious sensitivity of the human race when peer pressure pushes the buttons of psychological coercion!

Flawed dogmatic traditions impact politics, history, religion and science.

Authoritarian, lock-step conformity can infect classrooms.

When respected professors portray evolution as "fact," survival instinct kicks in motivating students to shrug and bite the hook. The most astute human minds are susceptible to cultural tradition and intense, repetitious touting of a party line.

© AlessandroZocc

Glasswing Butterfly

Totalitarian political regimes exploit this shortfall in human psyches.

Despite a century of frustrated misperception, the natural limits to biological change continue to stymie Darwin's dream.

"Nowhere was Darwin able to point to one bona fide case of natural selection having actually generated evolutionary change in nature." [19]

Confined to a whiff of time and a sliver of space, humans reach for evidence correlating our genesis with today and an infinity of tomorrows.

Evolution can't explain just how, when and where life first appeared on Planet Earth!

Evolution isn't happening now as a newsworthy current event; it hasn't happened in the recent past; it just never happened. The fictitious scenario features a counterfeit "science" that corrupts minds with a contagious intellectual virus.

More than an academic charade, belief in life's origin comes with consequences: purpose, quality of life and destiny impact all humans! Life is God's gift, for now and forever. To deliberately brush aside and ignore this miracle, demonstrates the incomprehensible arrogance of ignorance.

"…Evolution is the central most disorganizing, anti-intellectual anti-science principle that biologists have ever been dictatorially forced to learn to understand the world…it stands as the greatest scandal in science of the last 140 years." [20]

There is nothing in the living world that demonstrates evolution-in-action. Discontinuity reigns! Gaps dominate!

Evolution devotees may ignore the missing evidence, insisting the obvious gaps are exceptions to the "rule." Doubters are more likely to perceive the glaring exceptions tend to "eat up the rule."

For the unbiased observer, case closed!

For those clinging to a discredited idea, more bad news awaits!

"if any complex organ…could not possibly have been formed by numerous, successive, slight modification,"
Darwin predicted his theory would *"absolutely break down."* [21]

If Darwin's reasoning is to be taken seriously and ancient-time evolution is real, with life discounted as nothing more than an transitory accident, every fossil bone chip should represent a link in a continuous chain of organic life.

This is not the case!

Evidence of radical transformation in life formats by drawn-out whims of "gradualism" ranges from the suspiciously fragile to the non-existent.

Fast forward to century twenty-one and millions of fossil discoveries later, the *"most obvious and serious objection…against the theory,"* lingers in academic shadows, an unresolved nemesis!

With more than a couple billion bits and pieces of fossil bone fragments or imprints in the earth's crust having already surfaced, and an estimated 200,000,000 fossils displayed in world museums, it should be a slam dunk to match Darwin's dream with science reality---if Darwin's *"theory be true!"*

With only *"slight successive variations,"* and changes via *"short…slow steps:,"* fossil graveyards should overflow with millions of intermediates, transiting to radically new and diversified organic formats.

Instead, with the weight of evidence leaning heavily in the direction of nothingness, *"break down"* time knocks at logic's door! Those partially built, transiting plant and animals haven't been found.

They don't exist!

So where does that leave neo-Darwinism's molecule-to-man nonsense?

Numerous species have gone extinct---no question about it.

But as to that elusive *"organic chain"* of successor intermediates, the idea *"carries the same validity as a bedtime story—amusing, perhaps even instructive, but not scientific."* [22]

Where is the mythical Sherlock Holmes when he's needed to locate those still missing bits of ancient bones? Even the legendary Holmes couldn't pull mythical fossils out of sedimentary rock.

In the phrase made famous by Sherlock, its *"elementary my dear Watson."*

VI

Imagination Run Amok

Living Cell

"To get a cell by chance would require at least one hundred functional proteins to appear simultaneously in one place. That is one hundred simultaneous events each of an independent probability which could hardly be more than 10^{20}." [1]

Michael Denton

© Richard Fitzer

How did decorative feathers evolve from scales?
Cold-blooded reptiles don't make the cut as bird ancestors.

Two years before the publication of *The Origin of Species*, Charles Darwin wrote to Asa Gray admitting that in the absence of supporting evidence one's *"imagination must fill up the very wide blanks."* [2]

But even Darwin's innovative mind never came close to filling the gaps in his imaginings or grasping the magnificent beauty of the simplest living cell.

Night skies showcase more than ten million bright lights set against a dark field of infinite space. Invisible to the human eye, the simplest cell imaginable boasts a format of *"supreme technology and bewildering complexity"* built from *"about ten million atoms."* [3]

Primitive wisdom envisioned cells as little more than infinitesimal pieces of cytoplasm, *"a relatively disappointing spectacle appearing only as an ever-changing and apparently disordered pattern of blobs and particles."* [4]

Lacking access to electricity or the magic of the electron microscope, and presumably without knowledge of the genetic information revealed in Mendel's garden, Darwin crafted an idea that has been patched-up and presented to today's world as the Modern Synthetic Theory of Evolution.

Unaware of the typical prokaryote cell's complexity with its nucleus, DNA, protein and cell wall, Darwin imagined, incorrectly, that acquired physical traits could be transferred genetically to offspring.

Blind to his ignorance of cell composition, Darwin plunged ahead, publishing his imaginings in 1859, bypassing, as a *"far higher problem,"* the incongruity of original life emerging spontaneously from non-living matter.

This nagging lack of a rational naturalistic explanation for the origin of first life stretched beyond anything his mind could conjure-up. Evolution's premise, built on life from non-life, is more than mere dilemma---its scientifically catastrophic.

Did the distinctly different animal and plant kingdoms evolve from some accidental first life from inert matter in a *"warm little pond,"* or were these unique living formats designed and introduced during creation week, interdependent components of a global ecosystem?

The choice exits between fantasy and intelligent design! Darwin never claimed microbiologist credentials. But his instincts warned him of the unresolved design dilemma. Baffled by the lack of a viable scientific explanation for the sudden appearance of first life on Planet Earth, he struggled with the implications of his hypothesis.

What he admitted to be an *"utterly hopeless muddle"* remained unresolved throughout his lifetime. Eventually, twentieth century molecular reality exposed the lifeless roots of his conjectured *"tree of life."*

Before his 1882 death, Darwin confessed, *"…I am quite conscious that my speculations run beyond the bounds of true science."* [3]

When he died, the murky *"muddle"* had failed to inject even the hint of first life into the sterile stick of a figurative tree. Understandably, the naturalist ranked his grand scheme as *"…a mere rag of an hypothesis with as many flaw[s] & holes as sound parts."* [5]

Darwin's wanderings in a dark, intellectual wilderness, offered wide latitude for gross error. Aware of the primitive nature of scientific technology in the mid-nineteenth century, microbiologist Denton cuts Darwin some slack.

"There is little doubt that if this molecular evidence had been available one century ago it would have been seized upon with devastating effect by the opponents of evolution theory…and the idea of organic evolution might never have been accepted." [6]

Denton's vision of a living cell examined under the probing eye of an electron microscope opened vistas Darwin never imagined.

To evolve a penguin from a dinosaur presumably would require mutations evident in a series of millions of transitionals! Where are those fossil intermediates?

Throughout the first post-Darwin century, fossil bones anchored pre-history investigations, providing thin grist for Darwinian rhetoric. In one giant leap beyond nineteenth-century science's near-sighted focus on fossils, intricacies of the complex cell, previously unseen and unknown, took center stage under electron microscope scrutiny.

The new science of life pushed cell design, composition and maintenance front and center. Scrutinized under the probing eye of the electron microscope, the living cell emerged as a vibrant speck of organic life, more complex than any machinery yet designed by human intelligence.

What once had been dismissed as bland bits of protoplasm jumped out as genetic treasure troves. Molecular biologists pulled the curtain concealing the previously unseen to stare in wonder at a pulsating package of coordinated microscopic motors and machines driving life's core.

With the advent of molecular biology, the mysteries of life's origin shifted from obsession with miniscule bits and pieces of dry, fossil bones reconstructed in museums, to the microscopic mother lode of information---DNA's "Language of Life."

Once the electron microscope debuted with magnification capability of a million times, researchers devoured an unfolding panorama of knowledge.

Thanks to the power of the microscope, only four years after the 1925 *Scopes* trial, H. G. Wells and Julian S. Huxley spotted *"snakelike threads"* writhing *"slowly through the cell"* which they called *"mitochondria"* while noting they could see nothing *"inside the nucleus but a clear fluid."* [7]

This imperfect glimpse of a cell's inner workings proved but a hint of things to come! The simplest cell, throbs with life far more complex in structure and function than any mechanism yet conceived by humans.

Short decades after *Origin of Species* aroused debate, molecular biology surfaced raising eyebrows and temperatures. Investigation of previously invisible life forms turned naturalistic interpretations of the living cell's origin, structure and function, upside down.

The insightful discoveries revealed by molecular biology herald an unwelcome wake-up call for Darwin's fading dream!

Scrutinized under the probing eye of the electron microscope, the *"unparalleled complexity"* of a living cell stuns sophisticated observers.

A living cell consists of nitrogen, hydrogen, carbon and oxygen...it stores an encyclopedia of working knowledge, and it reproduces a copy of itself. The overwhelming discontinuity at nature's molecular level is highlighted in Michael Denton's *Evolution: A Theory in Crises!*

"To grasp the reality of life as it has been revealed by molecular biology, we must magnify a cell a thousand million times until it is 20 kilometers in diameter and resembles a giant airship large enough to cover a great city like London or New York.

"What we would then see would be an object of unparalleled complexity and adaptive design. On the surface of the cell we would see millions of openings, like the portholes of a vast spaceship, opening and closing to allow a continual stream of materials to flow in and out. If we were to enter one of these openings we would find ourselves in a world of supreme technology and bewildering complexity...

"The simplest of the functional components of the cell, the protein molecules, were astonishingly, complex pieces of molecular machinery, each one consisting of about 3,000 atoms...What we would be witnessing would be an object resembling an immense automated factory...larger than any city and carrying out almost as many unique functions as all the manufacturing activities of man on earth...a factory which would have one capacity not equaled in any of our own most advanced machines, for it would be capable of replicating its entire structure within a matter of a few hours." [8]

Beyond a pulsating pack of cytoplasm with a nucleus wrapped in a membrane, the cell contains machinery performing a multitude of actions every split second. Mitochondria, located within the cytoplasm, generate and store energy.

"Life is far more than chemicals, and building life immensely more complex than pasting carbon, hydrogen, oxygen, and nitrogen together in clever ways. Every coffin in the cemetery is filled with those same chemicals, but no one walks out in the morning." [9]

Not only is transitional continuity as predicted by evolution lacking in fossil fields, the molecular world of living cells shouts discontinuity!

"We now know not only of the existence of a break between the living and non-living world, but also that it represents the most dramatic and fundamental of all the discontinuities of nature...." [10]

"It is well established that the pattern of diversity at a molecular level conforms to a highly ordered hierarchic system. Each class at a molecular level is unique, isolated and unlinked by intermediates. Thus molecules, like fossils have failed to provide the elusive intermediates so long sought by evolutionary biology." [11]

No scientist has yet successfully synthesized life from inert non-life, nor explained the how, when and where first life originated by accident. *"…The chemical reactions required to form proteins and DNA do not occur readily. In fact, these products haven't appeared in any simulation experiment to date."* [12]

© Mayskyphoto

Biochemist Michael Behe, working with an assistant in the NIH lab *"…analyzed the supposed miracle of the first living cell coming into being by historical accident. 'What would you need?' they asked each other. 'You need a membrane, a power supply, and you need some genetic information. You need a replication system. And we kind of stopped and looked at each other. We said, 'Nah.'"* [13]

Francis Crick and James Dewey Watson earned their figurative spurs in the science hall of fame by uncovering DNA's delicate, double helix design in 1953.

Identification of DNA's incredible design clouded further the *"hopeless muddle"* of evolution's assumptive imaginings.

Nineteenth century "simple" cell dogma vanished, existing now only in the lexicon of obsolete science. Darwin's day academics would have been astounded had they been introduced to the invisible intricacies of one-celled Prokaryote organisms, wrapped in a membrane encasing a package of protein and a coded dose of DNA information.

A living cell's package of microscopic data reflects formulas as reliably predictable as any array of inorganic atoms depicted in the Periodic Table of the Elements.

A cell must absorb food, discard wastes, repair, replace, grow, possess regulatory mechanisms and reproduce---all functioning pursuant to a built-in information code. [14]

Evolution's postulated radical change of a life format to a new and entirely different plant or animal, can't occur, unless it happens here.

At an absolute minimum, a living cell requires a system of regulatory mechanisms; a constant supply of energy; an abundance of four nitrogenous bases; ribotide phosphates; twenty aminoacyl nucleotidates; deoxyribonucleic acid (DNA); DNA polymerase; and RNA polymerase. [15]

Ribosomes craft chains of proteins from a smorgasbord of twenty amino acids. Proteins don't form naturally from chemicals.

Never has *"one single functional protein molecule"* been discovered resulting from random chance processes! The twenty amino acid chains that make proteins can't order themselves.

Typically a protein consists of 500 amino acid chains. There are more than 30,000 distinct proteins. [16] It's a tall order to deliver 100,000 different proteins to the human body by random chance.

A smorgasbord of amino acid varieties provides the raw material from which proteins are built. *"…A protein may have many of each kind. A typical protein will have a few hundred amino acids…To make a protein that will do something useful, the cell has to get the right amino acids in the right order."* [17]

Protein molecules constitute *"…the simplest of the functional components of the cell."* Each of these molecular machines consists of *"about three thousand atoms arranged in highly organized 3-D spatial conformation…*

"The life of the cell depends on the integrated activities of… probably hundreds of thousands of different protein molecules…It would be a factory which would have one capacity not equaled in any of our own most advanced machines, for it would be capable of replicating its entire structure within a matter of a few hours." [18]

The key to proteins folding into specific, three-dimensional shapes is in the sequence and arrangement of amino acids.

After assembling in correct sequence, a protein's *"…long amino acid chain automatically folds into a specific stable 3D configuration…Particular protein functions depend on highly specific 3D shapes…Significant functional modification of a protein would require several simultaneous amino acid replacements of a relatively improbable nature."* [19]

Failure to fold correctly, risks disaster.

"The protein's most widespread role is as a catalyst in biochemical reactions, and in this role it is called an enzyme...Each reaction has its own enzyme…" which can *"speed up a reaction rate by at least a million…*

"An increase in rate by factors of ten billion to a hundred trillion are not uncommon…A factor of a hundred million means that what takes a thousandth of a second with the enzyme, would take about 3,000 years without it." [20]

Unlike the "which came first" chicken or egg quandary, the living cell débuted on the scene, hitting the ground running with all its proteins and DNA information in place and fully functional. There's no room for a multi-million-year time gap for either one to catch up to join with the other!

**Amino acids exist in non-living matter,
in both left-handed and right-handed formats,
living cells build exclusively from left-handed formats.**

Enzymes, protein forms within the cytoplasm, stand guard as catalysts, expediting the life processes of the cell. Without the lightening-like speeds introduced by enzymes, biochemical reactions would take so long, they could fail to function.

"Amino acid, when found in nonliving material…comes in two chemically equivalent forms. Half are right-handed and half are left-handed—mirror images of each other." [21]

Living cells build only from left-handed amino acids---never right-handed! The explanation continues to escape finite minds.

The left hand/right hand amino acid mystery doomed from the start, the human attempt to artificially structure life-from-non-life.

"…Amino acids produced in Miller's apparatus were both right-and left-handed amino acids, but right-handed amino acids are poisonous to living organisms. Right-handed amino acids render proteins nonfunctional." [22]

"Amino acids in life, including plants, animals, bacteria, molds, and even viruses, are essentially all left-handed. No known natural process can isolate either the left-handed or the right-handed variety.

"The mathematical probability that chance processes could produce merely one tiny protein molecule with only left-handed amino acids is virtually zero." [23]

Just as mysteriously, at death of an organic system, amino acids revert to inorganic matter's left-handed/right-handed status.

Even with a laboratory-created reducing atmosphere, without a valid technique for producing life-handed amino acids exclusively, the attempt to manufacture life artificially was doomed to certain failure.

The odds against the 2,000 enzymes essential to the simplest life form appearing spontaneously from inorganic matter, at one time and in one place, runs at something in the range of $10^{40,000}$ to one. [25]

Unlikelihood of such an event is astronomically beyond the 10^{50} impossible range.

By the year 2,000, the human genome, with its staggering three billion+ base pairs, had been deciphered. The complex system orchestrates 75-trillion human body cells.

Imagine each citizen of Planet Earth was given a Rubik Cube with the challenge to simultaneously resolve the puzzle in sixty seconds. Then be asked to correlate the colors again---also simultaneously and in sixty seconds! And again…and then again, etc!

© Galushko Sergey

You get the picture. Impossible, of course! But still more likely than a single, simple living cell creating itself from lifeless matter.

Beyond a pulsating pack of cytoplasm with a nucleus wrapped in a membrane, the cell contains machinery performing a multitude of actions every split second. Mitochondria, located within the cytoplasm, generate and store energy.

"Recently, Eugene Koonin and others tried to calculate the bare minimum required for a living cell, and came up with a result of 256 genes. But they were doubtful whether such a hypothetical bug could survive, because such an organism could barely repair DNA damage, could no longer fine-tune the ability of its remaining genes, would lack the ability to digest complex compounds, and would need a comprehensive supply of organic nutrients in its environment." [25]

"…200 million variations, ranging from microscopic red blood cells to long, skinny nerve cells that stretch from the base of the spine to the foot…Every cell contains an estimated one billion compounds…and among these compounds are five million different kinds of proteins…

"These compounds are highly variable in shape, size, electrical charge, and configuration; many can complete a function in a millionth of a second." [26]

Tissues, organs, and systems, all functioning in synchronized concert, are composed of living cells. *"…Organisms consist of a number of subsystems which are all co-adapted to react together in a coherent manner; molecules are assembled into multimolecular systems"* which combine *"into cells, cells into organs"* and organs into a *"complete organism."* [27]

Heart cells code for the heart, skin cells for the skin, and brain cells for the brain. What prevents a cell from crossing over and performing the wrong service for the wrong organ?

The cell's awesome complexity in the context of the fully functional human system might have brought assumption of its accidental origin to its knees but for evolution's entrenched bias.

Then where does that leave a postulate that a molecular composite of inanimate atoms might have assembled information for living cells from nothingness to accomplish what is beyond the competence of human thought?

Darwin's "tree of life" drawing fails to explain the source of its roots and mischaracterizes life's origin as an unexplained accident of nature, without direction or purpose! The mystery of a living cell creating itself accidentally from non-living matter, without the wisdom and power of an intelligent Designer, continues to baffle human minds.

Scrutinized by three-dimensional analysis, a self-created living cell remains spurious, uncorroborated conjecture despite attempts to award God's seal of approval. The complexity of the simplest known cell only complicates attempts to build a viable theory rooted in chance!

The theory's genealogy lacks authenticity since all life forms are alleged to be constantly on the move, transiting to something new and different. If taken literally, not only are all species subject to extinction and replacement, life in the here and now confronts only end-of-the-line nothingness.

Darwin didn't live long enough to hear the phrase "DNA" or to study the language of life it represented. Understanding the source of genetic information packed into each cell's DNA would ultimately push the mystery of life's origin by accident off the "Richter" scale of responsible scientific rhetoric.

As for critical genetic information, *"DNA (deoxyribonucleic acid) and RNA (ribonucleic acid) molecules, which are composed of complex arrays of amino acids and are the templates for all living organisms, have yet to be artificially created."* [28]

Evolutionists reject "Intelligent Design" theory, as less than true science on the grounds it is not readily falsifiable. The transparent incoherence of dismissing the design inference because it can't be readily tested, exposes evolution theory's own less-than-scientific fragile feet of clay.

David Coppedge summarizes a wide range of biological designs mimicked by industry, recognizing that *"in order to reverse engineer a system, it had to be engineered in the first place."* [29]

A Madagascar spider spins silk ten times stronger than Kevlar; jellyfish are being studied in order to build a *"better aquatic pump."* Design of the elephant's trunk is referenced for building a robotic arm. Shark skin provides the pattern for ship hulls and swimsuits. The optical flow of honeybee eyes provides guidelines for developing navigation systems capable of complex maneuvers.

The Modern Evolutionary Synthesis has yet to be verified in laboratory tests. More than that, Darwinian thought lacks the rationality inherent in the design inference.

"The Darwinian theory of descent has not a single fact to confirm it in the realm of nature. It is not the result of scientific research, but purely the product of imagination." [30]

Awed poet, Joyce Kilmer, describing the magnificent of the humble tree, wrote the famous lines, *"Poems are made by fools like me, but only God can make a tree."* [31]

The poet might have added, *"And only God can make a living cell!"*

VII

"Nonsense of a High Order"

DNA, Language of Life

"DNA contains the genetic blueprint of life…
It gives instructions to the rest of the cell to make proteins, and it
passes this same information on to the next generation … Without
DNA, living organisms cannot survive." [1]
Carl Werner

© Cynthia Kidwell

DNA's information calls the shots
dictating the species and the golden frog's color.

The DNA information carried by a single gene can no more be modified by mutation or human manipulation than medieval alchemists could make gold by mixing a recipe of other elements using abracadabra magic.

So where did a cell's information come from?

A blank computer disk offers a blank screen until some human delivers a digital code loaded with text or pictures. The shiny data bank lacks meaning or purpose without the deposit of a precisely coded message put in place by some intelligent source.

Once loaded, the coded message can be replicated *ad infinitum.*

Try playing a card game by shuffling a deck of blank cuts of white cardboard, each card lacking a number and a distinctive graphic print. No winners, no losers, no game---just a meaningless shuffle of nothingness.

Or try building a bridge without the data essential to design and transform raw steel into a graceful span of the Golden Gate.

A book with an attractive cover that binds a package of blank white paper would never make a best-seller list. Its the informative printed message that sells the book.

Matter, without information, lacks meaning! Matter, by itself, is abstract. Unproven theory epitomizes intellectual incoherence!

First graders, up to their ears in ABC's, recognize 26 letters in the English alphabet.

Kids learn to spell "cat" and "dog." By the time they reach high school, they have discovered the English language boasts a plethora of mind-bending combos of those 26 symbols---a vocabulary of significantly more than at least 200,000 English words with a capacity to communicate information on an encyclopedic scale. Impressive capacity, but not close to matching the information combos powering the "Language of Life."

© Joyce Boffert

**The Monarch butterfly's magical wings
demonstrate unique genetic information in action,
rather than centuries of mindless mutations.**

Monarchs summer in North America and then trek south annually, 2,000 miles to Mexico's warmth. Clouds of burnished gold wings, bound for their southland "resort," float at a gentle ten-miles-per-hour, typically covering fifty miles per day.

Monarchs find the identical winter home each year thanks to their own built-in GPS. The annual round-trip involves five generations---four that live only a month or so while the stalwarts that navigate the migration route to Mexico live nine months.

The source of the *"language of life,"*

pre-packed into a living cell, confounds imaginations.

Cells contain definitive information blueprints, unique and distinctive codes for every plant and animal species inhabiting today's Planet Earth.

Inorganic matter exists in the form of 100+ known atomic elements. The quantity of protons contained in the nucleus of an atom gives each element its distinguishing number.

Ten years after Darwin floated his ideas, Russian chemist, Dmitri Mendeleev arranged the 63 then-known elements in a chart of rows and columns dubbed a *Periodic Table of the Chemical Elements*. The chart confirms matter's intrinsic organization and precise atomic formulas.

Chemical recipes created from molecular mixes produce products that underwrite industry. Genetic information guarantees comparable precision! Cells of living organisms display elegant designs shaped by DNA's coded information built into every cell.

"What has been revealed as a result of the sequential comparisons of homologous proteins is an order as emphatic as that of the periodic table. Yet in the face of this extraordinary discovery the biological community seems content to offer explanations which are no more than apologetic tautologies." [2]

Michael Denton's, *Evolution: A Theory in Crises*, opened the door to a glimpse of molecular biology's avalanche of insight.

"Neither of the two fundamental axioms" of neo-Darwinism, *"…continuity of nature…linking all species together and ultimately leading back to a primeval cell"* and *"adaptive design…from a blind random process have been validated by one single empirical discovery or scientific advance since 1859."* [3]

Lacking viable explanation for both the origin of life from unintelligent non-life as well as the original source of DNA information that provides structure and direction to life, evolution theory hides behind unexplained assumption, glossing over Mother Nature's double-whammy!

Even if spontaneous generation had defied impossible odds by accidentally creating a "simple," living cell, it couldn't survive, much less reproduce itself, without its own package of DNA calling the shots.

Once the assumption virus invades reasoning, superstitious nonsense follows close behind.

With or without popular acclamation, assumption masquerading as science, doesn't evolve fact.

Its not plausible to contend a cell's information code originated spontaneously, by coincidence, from inert matter!

So where did that code of genetic information come from?

To speculate spontaneous generation triggered first life from inert matter strains credibility. Its off logic's wall to pony up unrealistic imaginings that offer no hard-core scientific explanation for the source of the genetic information vested in a living cell.

© NRPphoto

**Multiple protein families provide an array of
potential building blocks for specific life formats.
DNA exists *sui generis* (one of a kind).**

Its less than rational, and certainly unscientific, to suggest DNA's microscopic strings of pre-coded information appeared accidentally after millions of years of chaotic heating, cooling, and thawing mingled with rain torrents, lightening flashes, and wind gusts!

The nagging question persists: just where did that genetic information originate? Logic slinks out science's back door when the *"design inference"* is ignored.

Microscopic genetic information codes every cell of every living organism with a one-of-a-kind mark. Organic life kinds could not exist without a unique blueprint of genetic information housed in every living cell. *"Without DNA, living organisms cannot survive."* [1]

Every prototype plant and animal life on earth carries a staggering stash of unique genetic information, locked in place from the get-go. DNA data banks guarantee kaleidoscopic diversity in descendant generations but never radical change to a new and different organism.

DNA contains the master genetic blueprint for every

living organism with a precision reminiscent

of the Periodic Table of the Elements.

Tucked into the cell's nucleus are the strands of deoxyribonucleic acid containing the code of life for every organism---DNA for short. This compendium of living data is so miniscule, it can't be seen by the unassisted human eye!

The first living cell's DNA base pair could not exist without imprinted language of life instructions. A single cell's DNA comes loaded with information sufficient to fill 3,000 sets of printed encyclopedias.

"…The capacity of DNA to store information vastly exceeds that of any other known system; it is so efficient that all the information needed to specify an organism as complex as man weighs less than a few thousand millionths of a gram…Each gene is a sequence of DNA about one thousand nucleotides long." [5]

These ladder-like spirals are built on side rails with alternating molecules of phosphate and deoxyribose *"held together by 'rungs' called nucleotides (or bases) consisting of four specific chemical molecules: thymine (T), adenine (A), cytosine (C) and guanine (G)…Millions upon millions of nucleotides are known to exist in the nuclear DNA structure of living cells…"* [6]

Living organisms use these same nucleotides but in radically different formats just as cars and skyscrapers use steel but with different designs.

Nucleotides A and T bond together to form an AT or a TA base while G and C bond as GC or CG. *"A always bonds to T and the base letter G always bonds to C. Three contiguous letters on the DNA molecule (called a codon) instruct the cell to place one particular amino acid into a protein chain…RNA carries this template of the DNA and assembles proteins by attaching amino acids together in a chain."* [7]

"Nucleotides cannot be added at will; even if they did, they could not alight themselves in a meaningful sequence…..any physical change of any size, shape, or form, is strictly the result of purposeful alignment of billions of nucleotides. Nature or species do not have the capacity of rearranging them nor to add them." [8]

Sir Fred Hoyle scoffed at the concept of a genetic code emerging from some primordial organic soup by chance, branding it *"nonsense of a high order."* [9]

The genome's complexity defies explanation by accident.

Denial of DNA design, put in place at the command of an infinite intelligence, conjures up images of superstitious ancients, worshiping at the feet of inert matter, bowing in supplication before homemade idols of wood and stone.

The simplest known living cell, with its nearly 500 genes, is vastly more complex than any memorable musical composition or the most intricate machine created by humans!

A dragonfly sees through two eyes, each with 30,000 lenses. The bee gets by with 6,300. Genes dictate the difference.

The hexagonal honeycomb design in a bee hive utilizes a minimal amount of wax to store a maximum amount of honey. Where did the honeybee obtain the built-in design information that enhances honey's storage capacity?

Evidence of intelligent design pervades earth's life forms.

Potato bug eggs hatch in 7 days; canaries require 14; hen's eggs take 21 days; ducks and geese require 28; mallard ducks take 35; while 42 marks the timeline for the parrot and the ostrich. [10]

An elephant's four legs bend forward at the knees easing its rise to an upright position. In contrast, a horse first uses its front legs to stand while a cow starts with its hind legs.

Watermelons have an even number of stripes on the rind; oranges have an even number of segments; an ear of corn carries an even number of rows; and a stalk of wheat has an even number of grains.

And then there are banana bunches with an even number of bananas on the lowest row with each subsequent row decreasing by one, resulting in alternating even/odd rows for the bunch. [10]

Human genes design lachrymal glands essential to moisten eyes. Stag deer grow racks of antlers courtesy of gene code information. Thanks to distinctly unique genetic information, humans don't sprout horns on their heads and deer don't evolve tear ducts on their feet.

Just as fingerprints identify patterns unique to an individual, every body cell portrays the DNA sequence of a genome, unlikely to be precisely identical with that of another human. The individual master code that dictates eye color, sex, and the shape of the face, is usually one-of-a-kind.

DNA not only revolutionized investigation of life's origin but also emerged as a tool of the judicial system, shining light on crime detection and resolution. When DNA doesn't match crime scene evidence, convictions have been overturned, sometimes years-after-the-fact.

Internet entrepreneur Bill Gates, recognizes the limits of cyberspace language in contrast to the cell's ability to store and utilize living data.

"DNA is like a computer program but far more advanced than any software we've ever created." [11]

It's been calculated that if *"…all the copies of the DNA in all the cells of your body were straightened and laid end to end they would be about 50 billion kilometers long"* [12] and would stretch from the earth to beyond the Solar System.

DNA of self-replicating cells transcribes the code to messenger RNA which translates and transfers the data from the cell's nucleus to it's cytoplasm for protein synthesis.

Did the 50 billion kilometer length of microscopic strands of human DNA result from mega-millions of luck-of-the-draw mutations?

True science can't buy into random "chance" as a viable answer!

It has never been demonstrated that either the DNA or the RNA component of a living cell can arise spontaneously from non-living matter! Multiply nothingness by billions and the answer remains a vacuous zip.

Ribonucleic acid (RNA) acts as DNA's transfer agent taking the code from the nucleus to the cytoplasm.

With a single rail composed of phosphate and ribose molecules and rungs with uracil (U) instead of thymine (T), messenger RNA enters the nucleus at a moment when the DNA unwinds, reads and copies the information then delivers the code to a ribosome protein-producing factory.

"The order of dRNT-s of DNA determines its information content, provided that the rest of the cell's machinery is present." This refers to the *"complex apparatus which duplicates DNA, transcribes it to a readable message and then reads the message and produces a functioning protein…"* [13]

The complexity of a cell's information is compounded by the fact that not all genetic data resides in the cell's nucleus.

"…Mitochondria have their own DNA/RNA structure…a semi-independent order-giver in its own right, a computer sub-station so to speak…Human mitochondrial DNA has slightly more than 16 thousand nucleotides." [14]

If this infinitesimal information resource were placed in a teaspoon, together with DNA *"necessary to specify the design"* for all species of living organisms ever to have lived on the planet, *"there would still be room left for all the information in every book ever written."* [15]

Evolution's "RNA World Hypothesis" posits that RNA might have generated the evolution of a complete cell because it carries information and serves also as a catalyst for the manufacture of a cell's protein.

The "RNA World Hypothesis" is a dubious entry in the sweepstakes of the make-believe. Attempting to explain the origin of first life by accident, the idea credits *"11 small carbon molecules…that could have played a role in other chemical reactions that led to the development of such biomolecules as amino acids, lipids, sugars, and eventually some kind of genetic molecule such as RNA."* [16]

Little did those *"small carbon molecules"* realize the magnitude of the creative authority theoretically vested in their inorganic, atomic composition.

Living organisms consist of synergistic systems, organs, tissues, and thousands of miniscule cells in a single, functioning unit.

The language of life can't exist without proteins and proteins can't exist without the molecule that provides and stores instructional information.

RNA cannot survive on its own without being an essential part of the complete living organism---nor can it account for the source of its original genetic information!

DNA, RNA, proteins, amino acids, and cell membranes exist as a composite, mutually interdependent whole.

It's an all or nothing package deal.

Its axiomatic that a functional cell requires a precise dose of genetic directions to tell its amino acids and proteins just how to function as a one-of-a-kind, reproducing organism!

The genomes of different species show vast differences that defy bridging by simply reshuffling the card deck and re-dealing the hand. Genetic change will be confined to the potential combinations of the original 52 cards in the deck without adding new cards that deliver entirely new and different information.

© Randy Rowland

The genome of a Chimpanzee doesn't match the genome of a human. Nor is there a whit of evidence that an elephant, a giraffe, a butterfly, or a giant Sequoia could have descended from a common, single-cell ancestor.

Dean Overman discounts *"biochemical predestination"* in molecular genetics. Amino acids don't organize themselves without a genetic code orchestrating the process.

"The enormous information content of even the simplest living systems…cannot in our view be generated by what are often called 'natural' processes…

"For life to have originated on the Earth it would be necessary that quite explicit instruction should have been provided for its assembly…

"There is no way in which we can expect to avoid the need for information, no way in which we can simply get by with a bigger and better organic soup…" [17]

However imaginative, *The Origin of Species* could not have been written and published absent an intelligent act.

No evolutionist suggests Darwin's hypothesis evolved without thought-an idea stitched together accidentally by an obscure natural process.

Paradoxically, the publication at the core of the raging ruckus disputing intelligent cause in the origin of earth's life forms, is itself the product of intelligent cause.

It is intellectually incoherent to assert life and its built-in code of information appeared spontaneously, by chance, independent of intelligent cause. Like the machine with the sole function of turning itself off, human minds have proposed the inherently impossible---the creation of information from a non-intelligent source.

Lacking authentic scientific credentials for the origin of first life, belief in Darwin's unexplained *"far greater problem,"* becomes an act of faith---a secular religion.

The either/or choice seems obvious---faith in first life's origin by luck-of-the-draw chance or by an intelligent act?

A free society champions free speech in pursuit of true knowledge. Unless scrutinized critically at arms-length, evolution's subtle enticements can deceive the best and the brightest!

USA Today bought the bait in its August 9, 2005 edition, exalting evolution while taking a swipe at Intelligent Design. The feature endorsed evolution as *"…the cornerstone of modern biology. Though there are various 'missing links' in the evolutionary chain, it has never been refuted on a scientific basis."* [18]

Conversely, where's the evidence that evolution's *"working hypothesis"* has ever been proven *"on a scientific basis?"*

A team of intelligent media minds composed words and phrases, designed a layout, printed and distributed a message insisting that human brains originated by random chance from some unintelligent source---without design or designer.

**Science can't play the role of academic bookie offering
the public nothing better than impossible odds!**

So where's the "science" in a trail to the never was?

Canny Vegas gamblers wouldn't put their money on the line to back the impossible! Why would anyone bet their life on odds as unlikely as generating a stash of living information, by chance, from non-living matter?

"…Since most proteins are, on average, 300 amino acids long, the DNA needed to make just one protein would have to be approximately 900 letters long." And since the first living organism would have required 20 or so basic proteins to function, each with 900 letters of DNA, *"…18,000 letters of DNA would be needed to theoretically form the first single-cell organism."* [19]

Now factor in the length problem!

"Scientists have observed DNA strands forming naturally in the laboratory, but only in strands of up to 20 letters in length…

"After 20 letters of DNA come together, the DNA begins to break apart. Simply put, long strands of DNA (hundreds of tens of thousands of letters long) do not form naturally because chemical properties of DNA prevent this." [19]

Then there's the sequence problem!

"The 18,000 DNA letters have to be lined up in a particular order to call for particular amino acids in a particular set of proteins for life to begin…The odds of winning the national Powerball Lottery every day for 365 dars are 1/4,244 followed by 2,881 zeros…

"The chances of DNA forming spontaneously with the proper letter sequence" requires 10,837 zeros. [20]

Calculating the odds requires recognition of the number of genes to be reckoned with. *Mycoplasma genitalium*, with its 482 genes and 580,000 bases**,** may be the *"simplest known self-reproducing life form."* [21]

The genome of a mammal contains a string of *"from two to four million"* symbols *"…that would fill two thousand volumes---enough to take up a library shelf the length of a football field. All this is in the tiny chromosomes of each cell."* [22]

The mathematical odds against a single cell appearing by accident pushes computations over the edge. The already **impossible** eventually confronts the ultimate challenge of merging millions of cells into a cohesive living unit composed of a diverse variety of tissues, organs, and systems.

© Warren L. Johns

Roman aqueducts having weathered centuries, symbolize intelligent design, generated by bright human minds.

Based on mathematical probability factors alone, *"…any viable DNA strand having over 84 nucleotides cannot be the result of haphazard mutations. At that stage, the probabilities are 1 in 4.80 x 10^{50}…Mathematicians agree that any requisite number beyond 10^{50} has, statistically, a zero probability of occurrence…*

"Any species known to us, including the smallest single-cell bacteria, have enormously larger numbers of nucleotides than 100 or 1000…This means, that there is no mathematical probability whatever for any known species to have been the product of a random occurrence---random mutations…" [23]

Objectively analyzed from any angle, **impossible** spells **impossible!**

Molecular biology confirms discontinuity!

*"…The DNA molecule may be the one and only perfect solution to the twin
problems of information storage and duplication for self-replicating automata…*

*"It is astonishing…that this remarkable piece of machinery which possesses the
ultimate capacity to construct every living thing that ever existed on Earth, from a giant
redwood to the human brain, can construct all its own components in a matter of minutes
and weigh less than 10^{16} grams.*

This *"…is of the order of several thousand million-million times smaller than the
smallest piece of functional machinery ever constructed by man."* [24]

*"The difference in DNA between species resides almost strictly in the sequential
positioning of the nucleotide…No two individual plants or animals have DNA spirals
that are identical…"* [25]

Evolution theory notwithstanding, every cell of every living organism
codes with its distinctive mark.

"Bacterial genomes are very different from eukaryotic genomes [Bacteria are
Prokaryotes, single-cell life forms with an open internal cell structure while
Eukaryotes, also single celled, occupy multiple compartments] *in that they
usually do not possess 'exons' and 'introns' and do not have extensive scaffolding.
Bacteria even lack membrane-bound nuclei, so that their genes can be expressed much
more quickly than those of the eukaryotes…"* [26]

*"…Pure unguided random events cannot achieve any sort of interesting, or complex
end…The fact remains that nature has not been reduced to the continuum that the
Darwinian model demands, nor has the credibility of chance as the creative agent of life
been secured."* [27]

A congenitally flawed idea crafted from ambiguous abstraction lacks
substance. Molecular biology does nothing to demonstrate the genetic
continuity critical to the corroboration of evolution theory.

"At a molecular level…there is no trace of the evolutionary transition from fish to amphibian to reptile to mammal. So amphibia, always traditionally considered intermediate between fish and the other terrestrial vertebrates, are in molecular terms as far from fish as any group of reptiles or mammals." [28]

Serious scholars no longer cling to the fantasy that physical traits, acquired by use or disuse, can be transferred genetically.

Primitive minds investigating life sciences, studied what they could see—embryos and bones. Without access to the all-seeing eye of the electron microscope, similar appearance could be misinterpreted as relatedness.

Embryonic transformation from misunderstood "simple" cells to fully developed animals encouraged conjecture that cell transitions could lead eventually to most any complex plant or animal—given spin the bottle luck and mega years of evolutionary gestation.

Perceptions tainted by superstition, blew the same medieval hot air that floated spontaneous generation fallacy. The world might have been spared a library of specious speculations had molecular biology emerged as a nineteenth-century science.

It's universally recognized *"…physical growth is the result of a very specific sequence of hundreds of thousands of nucleotides in its own DNA"* which dictates the order giving sequence that *"seeps down to the growth mechanisms. It never acts in the reverse direction…"* [29] Lacking evidence of a convincing pattern of transitional life forms after millions of fossil discoveries, evolution theory lives or dies in DNA's molecular world.

But the microscope doesn't oblige! So back to the drawing board!

Just where did that information stashed in DNA actually come from?

Rational thought declares an intelligent source to be imperative!

Certainly DNA didn't drift in riding a drifting, spontaneous fog playing roulette with nature's chemistry set.

Michael Denton confirms the inability of chance to produce change necessary to close the massive gaps existing at the molecular level.

"No evolutionary biologist has ever produced any quantitative proof that the designs of nature are in fact within the reach of chance…It is surely a little premature to claim that random processes could have assembled mosquitoes and elephants when we still have to determine the actual probability of the discovery by chance of one single functional protein molecule." [30]

Evolution theory hangs by a tattered thread, trapped by the litany of unproven assumptions its sponsors spin to the public. With a giant zip on the science scoreboard how does the idea survive?

What keeps a bankrupt franchise in business?

Evolution's devalued life concept exists exclusively on paper. Little else remains for the fragments of Darwin's dream but an overdue eulogy. The core question deserves repetition: *just where did the information, pre-loaded into the cell's DNA, come from?*

Objectivity directs the intellect to a *design inference* answer!

Without evidence as to the why, when, where, what and how DNA's "language of life" managed to find its way into the heart of a first living cell, what is the basis for claiming all descendant life forms ended up on a "tree of life" that evolved itself from nothingness en route to eventual oblivion?

In a vain quest for transitionals, evolution theory lost its academic battle in fossil cemetery fields and molecular biology laboratories. Instead of offering salvation for a decrepit idea, molecular discontinuity is writing the theory's overdue obituary.

Early in the twentieth century, the ball was passed along to mutations in a last, desperate effort to salvage flawed fiction.

"Refuse Material of Nature's Workshop"

Mutations

"No matter how numerous they may be, mutations
do not produce any kind of evolution…There is no law against day dreaming,
but science must not indulge in it." [1]

Pierre-Paul Grassé

© Ng Wei Keon

Does a proud peacock owe its beauty to mutations?

When Darwin first floated his philosophy, he was as ignorant of DNA as he was of a flash drive for storing computer information. But once having shared his conjecture with the public, he may have felt compelled to offer an explanation as just how those incremental transitions occurred.

Already out on a limb of his *tree of life*, he latched on to the mystical thought that perhaps physical change, activated by the use or disuse of a body part, could be passed along to a descendant.

Darwin differentiated between *somatic* and *germ* cells by crediting what he labeled *gemmules* as the mechanism that would preserve and pass along acquired physical traits from somatic cells to germ cells and then on to descendant generations. Bit-by-bit, gradual changes supposedly accumulated en route to an entirely new and different life kind.

The hopeful guru seized upon the towering neck of the giraffe, a top ten attraction at animal zoos, as prime evidence of the process.

Fairy tale logic manufactured the fiction that when persistent droughts dried the fields of grass grazed by giraffe ancestors, survival hinged on stretching their necks toward the sky to nibble tree leaves.

Theoretically, the animal's bone and muscle structure managed to preserve stretches of the neck and forward each miniscule change to the next generation, thanks to those mysterious *gemmules*.

The never-was *gemmule* is defined as *"a hypothetical particle of heredity postulated to be the mediating factor in the production of new cells in the theory of pangenesis."* [2]

Accrued over time, these modest adaptations supposedly evolved the long-necked giraffe that populates African plains.

Never observed in nature, the far-fetched myth made no sense.

© Chuck Nelson

Darwin believed the fiction that acquired physical traits could be passed to future generations genetically. Today's evolutionists substitute mutations as the component for radical change.

Darwin's *gemmules* existed only in his mind. It would be well into the early twentieth century before the world would catch-on to the wishful thinking.

As to the neck of an 18 foot-tall giraffe with a heart 2 ½ feet long, mutations contribute nothing more to explain its origin than do make-believe *gemmules*.

The heart powerful enough to pump blood up the extended neck to the brain is also powerful enough *"to burst the blood vessels of its brain"* when it reaches down for a drink of water. But when the giraffe bends down, *"a protective mechanism"* kicks in causing *"valves in the arteries in its neck"* to begin to close. [3]

Why didn't other animals, such as zebras, evolve longer necks as well? And why couldn't a human muscle builder pass along to his children, a set of enhanced muscles, built by exercise?

Embryonic transformation from misunderstood cells to a fully developed animal encouraged conjecture that comparable transformations of any cell could lead eventually to most any complex creature—given spin the bottle luck and mega years of evolutionary gestation.

Theories of origin came cluttered with the same medieval hot air that floated spontaneous generation fallacy. Oblivious to genes and DNA, evolution's pioneer guru went looking in the wrong direction for theory confirmation. He relied on the fossil record, despite its paucity!

Darwin's optimism that evidence from fossil fields would reveal a continuous chain of organic life linking simple to complex, lay in disarray.

The eloquence filtering through the stentorian tones of a persuasive English accent proved vulnerable to superstition, tradition and personal bias prior to laboratory access to tools of modern technology.

Knowledge as to ancestry and origins lay obscured in data invisible to the naked eye. Not until the advent of molecular biology did doors swing open, inviting comprehension. The world might have been spared specious speculations had molecular biology been a nineteenth-century science.

Inevitably, belief that acquired physical traits could be passed on to offspring proved gross error. Either Darwin hadn't heard or had ignored the insightful news from Mendel's garden, released in 1865!

When the twentieth century dawned, Gregor Mendel's garden set the stage for a direct assault on the Darwinian notion that so intrigued late nineteenth century academia.

By the time Mendel's Law surfaced, Darwin's theoretical reliance on acquired physical traits as the means for blazing trails to new and entirely different species had lost credibility.

Gregor Mendel (1822-1884), an obscure Czech monk and Darwin younger contemporary, discovered fundamental genetic principles in the quietude of a monastery garden shortly after the English naturalist went public with his postulated surmisings.

Mendel's experiment revolutionized knowledge of the mechanism of inheritance. He bred several generations of garden peas focusing on the plant's seven basic characteristics. What he discovered about hybrids clashed head-on with Darwinian conjecture.

Mendel read his findings before the *Brünn Society for the Study of Natural Science* in 1865. His eyebrow raising report appeared in the Society's *Journal* in 1866, warranting distribution to 120 libraries, including some in England and eleven in the United States.

Mendel's scholarship, otherwise ignored or overlooked at the time, earned reference in the *Encyclopedia Britannica's* 1892 edition.[4] The landmark findings, obscure and virtually unnoticed, eventually were translated into English and published in 1900.

Pushing the public release was 39-year-old British biologist, William Bateson (1861-1926), an admirer of Mendel's scholarship, founder of the science of genetics and eventual president of the British Association for the Advancement of Science.

Publication rocked intellectual circles!

There is speculation as to whether Darwin's theory of evolution would have "evolved" and seen the light of day had he studied Mendel's findings. Bateson, a British biologist, expressed doubts.

"Darwin would never have written the Origin of Species if he had known Mendel's work." [5]

Maybe "yes," maybe "no," but we'll never know!

What we do know is that modern cell knowledge, with discovery of its DNA strands, correlates with Mendel's law of heredity and does nothing to corroborate Darwin's theory. Darwin's co-evolutionist, Alfred Russell Wallace, viewed Mendel's findings with alarm, posing a threat to evolution.

"On the general relation of Mendelism to evolution, I have come to a very definite conclusion. That is, that it is really antagonistic to evolution." [6]

Mendel's pioneer research launched the science of genetics confirming life forms function through precisely expressed information codes with predictable results. His law of genetic inheritance blew the cover off empty rhetoric embellishing make-believe. The discovery opened doors to genes, chromosomes, and DNA—the master genetic templates that determine and distinguish life prototypes.

"Cell biologists identified chromosomes as the carriers of Mendel's heredity factors, and in 1909 Wilhelm Johanssen named them 'genes.'" [7]

What Mendel did with garden peas, Hugo deVries confirmed with the primrose. He reported flowers rising *"...suddenly, spontaneously, by steps, by jumps. They jumped out among the offspring."* DeVries described these hybrid variables as *"new species"* which he labeled *"mutations."*

The word was out! Given time, *"mutations"* were incorporated as key componentsof evolution's iconic lexicon. [8]

In 1908, Alfred Russell Wallace (1823-1913), sensed a threat to evolution theory and dismissed Mendel's discovery, discounting *mutations* as minimally significant.

"As playing any essential part in the scheme of organic development, the phenomena seem to me to be of the very slightest importance. They arise out of what are essentially abnormalities, whether called varieties, 'mutations,' or sports." [9]

He saw *"...their extinction under natural conditions more certain and more rapid, thus preventing the injurious effects that might result from their competing with the normal form while undergoing slow adaptive modification...*

"Any species which gave birth to a large number of such abnormal and unchangeable individuals would be so hampered by them whenever adaptive modification became necessary that the whole species might be in danger of extinction."

Wallace scorned these abnormalities as *"refuse material of nature's workshop, as proved by the fact that none of them ever maintain themselves in a state of nature."* [9]

Presumably unforeseen by Wallace, his dismissal of mutations as mere *"refuse material,"* ran diametrically counter to subsequently postulated *Modern Synthetic Theory of Evolution*! He didn't live long enough to learn that what he described as *"abnormalities"* would be seized upon eventually as the holy grail used in an attempt to salvage a crumbling tradition.

Wallace was not the only influential evolutionist reluctant to jump for joy when confronted with Mendel's law of genetic inheritance.

No evidence has been presented that confirms a cardinal's elegant beauty, color and design resulted from a series of accidental mutations.

The mutation news shook the faith of others with seismic impact.

Princeton's Prof. Scott complained the findings *"…rendered but little assistance in making the evolution process more intelligent, but instead of removing difficulties have rather increased them."* [10]

A shaken Chair of Evolution professor for the University of Paris, voiced alarm to a 1916 Harvard audience.

"It comes to pass that some biologists of the greatest authority in the study of Mendelian principles of heredity are led to the expression of ideas which would almost take us back to creationism…

"The data of Mendelism embarrasses us quite considerably." [11]

Zoology Professor, E.W. McBride, reminisced for *Science Progress* the year of the 1925 Scopes Trial. He recognized Mendel's Law as a potential wet blanket cast over evolution theory after first being greeted with *"enthusiasm."*

His *"cul-de-sac"* analysis summarized the dilemma.

"We thought at last the key to evolution had been discovered.

"But as our knowledge of the facts grew, the difficulty of using Mendelian phenomena to explain evolution became apparent, and this early hope sickened and died. The way that Mendel pointed seemed to lead into a cul-de-sac." [12]

Its more than a quantum leap to suggest a string of DNA from human cells stretching 50 billion kilometers into space could be built by multimillions of random chance quirks of nature flowing seamlessly from earth's first ever life form.

The Mendel effect reverberated far beyond garden peas. Evolutionists recognized the sheer absurdity of *gemmule* poppycock.

The daunting revisionism task was handed to Wallace's *"refuse material."*

Mendel's law caught the attention of scholars and crept into science vocabulary early in the 20th century.

The stage beckoned for a theory upgrade that would rally the faithful. Convinced acquired physical traits could never account for transitional genetic changes, concerned evolutionists scrambled for alternatives for a discredited theory facing collapse on academic ropes.

And not a moment too soon! Crescendos of doubting voices could be heard reacting negatively to eroding credibility.

"[T]he theory suffers from grave defects, which are becoming more and more apparent as time advances. It can no longer square with practical scientific knowledge, nor does it suffice for our theoretical grasp of the facts…

"No one can demonstrate that the limits of a species have ever been passed. These are the Rubicons, which evolutionists cannot cross. Darwin ransacked other spheres of practical research work for ideas…but his whole resulting scheme remains, to this day, foreign to scientifically established zoology, since actual changes of species by such means are still unknown." [13]

Evolution has shown innovative flexibility when its survival is at risk. During the 1859 to 1872 interval, Darwin himself, repeatedly modified his *Origin of Species* manuscript by increasing the original 150,000 word text to 190,000 words. [14]

Responding to Mendel's law, evolutionists wracked collective brains for some fashionable alternative to genetic transition achieved by acquired physical characteristics. Rather than a once-perceived troublesome nemesis to a flawed idea, mutations emerged as the cure-all designed to redeem an endangered idea from the slagheap of discredited "science."

Theodosius Dobzhansky floated the mutation life preserver in 1937 celebrating this very same *"refuse material"* as evolution's *"raw material."* He postulated *"mutations and chromosomal changes… constantly and unremittingly supply the raw materials for evolution."* [15]

Big problem here!

Not only do mutations corrupt genetic codes, but they can kill!

In 1941, evolutionist movers and shakers put their heads together in an effort to reinvent Darwinian thought by blazing a trail through the quick sands of genetic mistakes.

Fleeing the discredited *gemmule* pathway to nowhere, revisionists seized Dobzhansky's straw in the wind, replacing the *gemmule* travesty by passing the mantel to *mutations* as the *raw material* enabling natural selection to do its thing. This legerdemain turned the looming threat of the *mutation* negative into a newly minted positive.

Designating *mutations* to serve as the *raw materials* for evolution seems as fanciful as envisioning the equivalent of a thundering Niagara Falls spraying its mists in a Mojave Desert mirage.

Duly revived, the now updated dogma was awarded the lofty title, *Modern Synthetic Theory of Evolution*, and reintroduced to the world with fanfare's bells and whistles.

The sleight-of-hand reinvented *"refuse material"* to read *"raw material"* that would supposedly partner with *natural selection*, thereby rewriting the anchor mechanism powering Darwin's dream.

Faithful Darwinists, symbolically walking in "tall cotton," met in Chicago in 1959 to celebrate the centennial anniversary of the publication of *The Origin of Species*.

Sir Julian Huxley, grandson of the 19th century Huxley who ran public interference for Darwin's original thought, made no apologies for the patchwork massaging of Darwin's grossly incorrect handiwork.

Waxing eloquent, Sir Julian testified to evolution's authenticity!

He assured celebrants, *Darwin's theory…is no longer a theory but a fact…We are no longer having to bother about establishing the fact of evolution…"* [16]

In a verbal swipe, he ruled out *"either need or room for the supernatural…"* [16]

Despite the exuberant chest thumping, the revised synthetic theory drew cheers mixed with some jeers. Early in this century, more than 500 scientists signed-on to a declaration expressing reservations.

"We are skeptical of claims for the ability of random mutation and natural selection to account for the complexity of life. Careful examination of the evidence for Darwinian theory should be encouraged." [17]

Suggesting mutations evolve new life kinds makes no more sense than claiming disease promotes good health.

Not only are mutations virtually synonymous with genetic degradation, no valid evidence demonstrates that mutations have ever successfully provided the raw information essential to transit one living organism to some new and different life kind!

Redundant clichés citing finch beaks, fruit flies, and bacteria modifications as evidence of evolution, confirm only the versatility potential built into a preexisting genetic code. Genetic chasms, separating distinctly different kinds of organisms, have yet to be bridged by the conjectured mutation/natural selection combo!

No single mutation or series of mutations, collaborating with natural selection, can activate random chance change to a new life kind without the addition of new genetic information.

Mutations paired with natural selection may shift a descendant kind of life laterally or down but never vertically up the genetic staircase to an entirely different kind of *family, order, class* or *phyla.*

"Mutations take place, but they are either reversible, deteriorative, or neutral…If one must depend on mutation and natural selection to produce new species---let alone, new families, orders and phyla as evolutionists assume, then not even billions of years would suffice." [18]

© Teli Képék

Incapable of introducing new information, mutations degrade overall fitness, typically impairing what's already there! Rather than a silver bullet to salvage a bankrupt theory, mutations have yet to find a niche in the pantheon of respectable science. Mutations typically degrade fitness!

Natural selection is genuine! But don't hold your breath expecting evolutionary change if natural selection relies on mutations. Assuming mutations provide the *"raw material"* for natural selection to create new and different living organisms makes no more sense than believing an improved computer evolves from a virus-infected hard drive.

Natural selection works counter to the other half of evolution's equation by screening out harmful mutations! Thanks to repair enzymes, mutations are subject to an organism's self-correcting mechanism. If the repair system itself mutates, the organism's survival could be jeopardized. Mutations are much too slow to accomplish evolution's conjectured mission.

"Natural selection can serve only to 'weed out' those mutations that are harmful, at best preserving the 'status quo.' " [20]

"Natural selection can act only on those biologic properties that already exist; it cannot create properties in order to meet adaptational needs." [21]

"No one has ever produced a species by mechanisms of natural selection." [22]

Natural selection alone, offers nothing more than tautology's circular reasoning.

"…The Darwinian theory of natural selection, whether or not coupled with Mendelism, is false." [23]

"The genome is fragile, subject to environmental insults (radiation, oxidation etc.) Extensive repair systems are guarding the integrity of the genome second-by-second. Without their work, life would become extinct." [24]

Genetic code corruption by mutation doesn't provide evolution's *"raw material"* but does contribute to deterioration of the genome.

Reinventing evolution by reliance on mutations as the theory's *"raw materials"* is fanciful. Desert mirages don't deliver sparkling water! A single mathematical miscalculation can disrupt a rocket's moon shot. Mistakes in pharmaceutical products can kill.

Ultraviolet light, nuclear radiation, and those pesky virus parasites rank as prime candidates for mutating genetic codes.

Just as electronic viruses wreak computer havoc, organic viruses lead a parade of enemies to life. Living, parasite viruses inflict permanent damage on a genome by corrupting genes.

Beautiful, vivacious, four-year-old Kylie McPeak appeared to be every parent's dream of an ideal daughter.

But then, her mom sensed a "feeling" that something was wrong.

The caring mom had reasons to worry!

By the time Kylie was six, the twitching became an uncontrollable writhing and her sweet child's voice began to quaver. Stumped local doctors encouraged the parents to present Kylie's case to the National Institute of Health's Undiagnosed Diseases Program. [25]

Baffled for a time, UDH eventually spotted the culprit---a mutated gene present on human chromosome 6 that codes for the protein laforin, usually a heart-wrenching death sentence for the child victim.

The genetic mutation that triggers Lafora disease in kids is a poor candidate as evidence that mutations provide evolution's *"raw material!"*

So what's going on here? Where is that new genetic information essential for natural selection to select in order to evolve?

The virus culprit lurks front-and-center as a prime villain! The end of the twentieth-century had identified a growing roster of serious mutation-caused genetic disorders, including the dreaded HIV curse.

Evidence that virus parasites appear to be anything but life-friendly continues to pile up. In recent years, millions of honeybees in the United States have died, mystifying science until 2010. It is now believed the twin culprits may be the insect iridescent virus attacking bees inflicted with the Nosema ceranae fungus [26]

The havoc wreaking CCD (Colony Collapse Disorder) contributes nothing positive in support of evolutionary change!

HIV, the human immunodeficiency virus that causes AIDS, is a retrovirus. Retroviruses are viruses composed of RNA.

© mirvev

Colony Collapse Disorder devastates the honeybee population when the insect iridescent virus combines with the fungus Nosema ceranae to infect a hive. A virus, with a fungus, put honeybees at risk.

They contain an enzyme that gives them the unique property of transcribing RNA (their RNA) into DNA. The retroviral DNA can then integrate into the chromosomal DNA of the host cell to be expressed there.

The origin of a virus mystifies. The havoc it inflicts is clear!

Its axiomatic: lacking ability to reproduce itself without access to a living, host cell, a virus doesn't qualify as a candidate for first life on Planet Earth.

Since the parasite cannot reproduce without latching onto a pre-existing "host cell," the how, when, and where the insidious "mini-monster" first intruded on pre-existing life confounds.

The negative impact of the virus far outweighs its miniscule size. It can mutate both DNA and RNA!

"The viral DNA or RNA then takes control of the cell's functions, including reproduction. The host cell has no choice but to make copies of the virus until the cell explodes, sending hundreds of viruses out to other healthy cells." [27]

Non-mutational changes in the appearance of a life form that can modify physical appearance---non-genetic factors that don't modify underlying DNA sequence---may change gene expression that can alter appearance.

"The modification of epigenetic features associated with a region of DNA allows organisms, on a multigenerational time scale, to switch between phenotypes that express and repress that particular gene...

"...Most of these multigenerational epigenetic traits are gradually lost over several generations...

"When the DNA sequence of the region is not mutated, this change is reversible." [28]

Reversible modifications of gene expression do not change the underlying DNA sequence of an organism.

Ongoing discoveries in the molecular world suggest mutations may prove to be Mendel's monkey wrench, undermining evolution theory.

IX

Gregor Mendel's Monkey Wrench

Darwin's Nemesis

"Major functional disorders in humans, animals and plants are caused by the loss or displacement of a single DNA molecule, or even a single nucleotide within that molecule...There is no evidence for beneficial...genetic mutations." [1]

Richard Milton

Wikipedia

Gregor Johann Mendel's (1822-1884) discovery of the Law of Genetics threw a figurative monkey wrench into evolution's iconic mechanism.

Mendel's Law of Genetics demonstrates precision in the world of organism comparable to the Table of the Elements in the inorganic.

"When a single mutation occurs in a single newborn, even if it is a favorable mutation, there is a fair probability that it will not be presented in the next generation because its single carrier may not, by chance, pass it on to its few offspring." [2]

Change in a single nucleotide would be the smallest possible modification of a genome. One evolutionist guesstimated that as few as 500 mutations could evolve a new species.

Mathematic odds challenge that possibility.

"It's a matter of chance that a mutant survives. It might spread through the population and take it over, but more likely it will just vanish…even good mutations are likely to disappear from the population.

"The chance of 500 of these steps succeeding is 1/300,000 multiplied by itself 500 times. The odds against that happening are about $3.6 \times 10^{2,738}$ to one, or the chance of it happening is about $2.7 \times 10^{-2,739}$…It's more than 2,000 orders of magnitude smaller than…impossible." [3]

Computations of the likelihood for any single event beyond 10^{50} chances qualifies as impossible unless absolutism is willing to concede the intrusion of a miracle.

Impossible trumps improbable.

What are the odds of six billion human dice throwers coming up with identical number sets at the precise nanosecond of time in 4.6 billion years, assuming each player could live so long? The odds stretch plausibility!

© Chuck Nelson

Suggesting a butterfly's delicate design resulted from mutations, not only bends but breaks Mendel's Law of Genetics.

No matter how many billions of chances and multi-millions of years allocated, the chance of those 500 allegedly beneficial mutation steps succeeding continues impossible ad infinitum.

Even an artfully loaded pair of dice will never roll a number higher than 12 no matter how many tosses.

Darwin's co-evolutionist, Alfred Russell Wallace, viewed Mendel's findings as a threat to evolution theory.

"On the general relation of Mendelism to evolution, I have come to a very definite conclusion. That is, that it is really antagonistic to evolution." [4]

He had reason for concern!

As to the tired cliché that mutations benefit bacteria by building immunity to antibiotics such as streptomycin, reality indicates *"the mutation reduces the specificity of the ribosome protein, and that means losing genetic information"* with *"a loss of sensitivity."* Despite some *"selective value,"* this mutation *"decreases rather than increases genetic information."* [5]

"...Antibiotic resistance is the result of loss of a protein, loss of the binding capacity of a protein, or the loss of a transporting system...It's a loss of something...

"If you're removing a transport protein to eliminate the bacteria's sensitivity to antibiotics, then how is that explaining common descent by modification...?" [6]

"There aren't any known, clear, examples of a mutation that has added information." Instead, mutations lead *"to a loss of sensitivity to the drug...the effect is heritable, and a whole strain of resistant bacteria can arise from the mutation...A change in one of its proteins is then likely to degrade the organism...Information cannot be built up by mutations that lose it."* [7]

"For information to build up in living organisms, it must be created somewhere... Although in some special cases a loss of information can lead to an advantage for the organism, the large-scale evolution for which the NDT [neo-Darwinian Theory] is supposed to account cannot be based on such mutations." [8]

"...Most mutations which cause changes in the amino acid sequence of proteins tend to damage function to a greater or lesser degree...most of the amino acids in the centre of the protein cannot be changed without having drastic deleterious effects on the stability and function of the molecule." [9]

"There are all kinds of mutations that eliminate proteins. They may eliminate transport protein, an enzyme, the action of an enzyme, or regulatory systems." [10]

Mutations are *"not making new transport proteins! They're not making new regulatory systems! Antibiotic resistance is an example ...*

"Every time you read about antibiotic resistance...they're going to talk about this as...an absolute example of evolution.

"There's no mutation that gives them [evolutionists] what is necessary for common descent with modifications." [10]

One of Darwin's false assumptions *"was that natural selection had a building or creating capacity, and it doesn't."* It removes information.

"Every molecular example of a mutation we currently have fails to provide a mechanism than can account for the origin of any genetic activity or function." [10]

Since 1990, *"discoveries of heart-handicapping mutations have been pouring out of numerous labs at an ever-increasing rate, yielding more than 100 mutations in more than a dozen genes."* [11]

"A genetic polymorphism called 11307K in either of...two APC genes doubles the risk of colon cancer." [12]

Crippling mutations respect no human!

Mutations unload a bleak litany of physical flaws and debilitating diseases on the genome!

Harmful gene mutations have plagued humans with a list of 4,500 already identified bad results—and still counting.

**Despite today's awesome diversity of bird species,
some Darwinian loyalists insist that birds
"evolved" from mutated dinosaur genes.**

A once monster of a man, now twisted and bent almost in half, totters feebly, aided by a walker. Parkinson's disease, the insidious scourge, laid low the robust physique of the former football star. A mutated gene stands accused as the culprit! [13]

Progressive myoclonus epilepsy is caused by a gene mutation on chromosome 21. Treacher Callins Syndrome, hemochromatosis, is linked to a defective gene on chromosome 6.

A caring, elementary school teacher, was forced to cope with Acromegaly's brutal disfigurement and the bullying taunts of the mean spirited, inflicted by an acquired, genetic mutation.

A gene mutation on chromosome 5 generates deformities of the face, ears, down-slanting eyes, and deafness.

A chromosome 9 gene mutation causes skin cancer. A mutated gene on chromosome 16 is tied to fanconi anemia, affecting children who rarely live past their sixteenth birthday.

A gene missing from chromosome 7 causes Williams Syndrome.

Anhidrotic ectodermal dysplasia, resulting from a mutation on the X chromosome, can afflict victims with baldness, loss of teeth, or deprive them of ability to sweat. [6][14]

"…Not one mutation that increased the efficiency of a genetically coded human protein has been found. Instead of a 'blind watchmaker,' the mutations behave like a 'blind gunman,' a destroyer who shoots his deadly 'bullets' randomly into beautifully designed models of living molecular machinery. Sometimes the 'bullets' only cause minor damage; sometimes they maim and cripple; sometimes they kill." [7][15]

There are *"genetic flaws that make people fat…"* [16]

Werner's syndrome results from a mutated site on human chromosome 8 causing victims to *"age prematurely fast and usually die before they reach 50."* [17]

Yet another genetic mutation *"…causes children to die of old age…Children with Hutchinson-Gilford progeria syndrome age at a rate five to 10 times faster than normal. They lose their hair, their skin wrinkles, and they die of arteriosclerosis, or hardening of the arteries, by their early teens."*

The defect is in *"…a gene that controls the structure of the nucleus…"* [18]

Progressive blindness described as retinitis pigmentosa is a disorder linked to a gene from the X chromosome.

"Best's macular dystrophy…destroys the part of the retina responsible for the sharpest vision 'has been linked to mutations' in the gene now called bestrophin." [19]

Spontaneous blood clots can form with the power to cause sudden death where a *"patient with the disorder has inherited at least one defective gene encoding protein C."* [20]

A mutated gene on chromosome 11 contributes to inherited hearing impairment. [21]   Then there's the smorgasbord of birth defects caused by mutated genes: muscular dystrophy, spinabifida, cystic fibrosis, Huntington's Disease, hermachromatosis, and Down's Syndrome.

Leukemia may result when a piece of chromosome translocates to another chromosome in the midst of cell division.

Regarding humans, researchers *"calculated an unusually high rate of 4.2 mutations per generation, of which 1.6 diminish the fitness of the species...the species must survive in part because people who have accumulated dangerous mutations are least likely to successfully have children...*

"The human reproductive strategy helps purge harmful mutations in batches...they mix their genes with another's, and presumably some of the worst defects aren't passed along. That wouldn't happen if humans reproduced asexually." [22]

Mutated human genes labeled APOE, CLU, and PICALM have been targeted as likely culprits inflicting Alzheimer's disease on the lives of five million Americans. [23]

"Meat contaminated by a virus that can be lethal is being sold to consumers -- creating a public health threat that has largely flown under the the radar due to powerful industry interests and lax accountability at the federal agency in charge of ensuring food safety, according to recent studies and a prominent investigative journalist.

"'It makes salmonella look like a picnic,' is how David Kirby, an investigative journalist who has written about MRSA, a life-threatening pathogen, described it in an interview with Consumer Ally.

"MRSA (Methicillin-resistant Staphylococcus aureus) is an antibiotic-resistant staph infection that kills about 20,000 Americans -- more than the number of people who die from AIDS -- each year." [24]

DNA is so delicate that one report suggests *"Every exposure to tobacco, from occasional smoking or secondhand smoke, can damage DNA in ways that lead to cancer."* [25]

Inevitably, evolution's mutation "cure-all" itself will suffer further from still to come discoveries, certain to expose virus-caused-mutations for the culprits they are---debilitating genetic mistakes.

The glaring exception to the universal commonality of the science of order and the mathematically real is evolution's mutant "science" where the logic of the measurable equation is discarded in favor of mutation's assumptive myth.

Evolution ignores the real and postulates the imaginary. Empty rhetoric lacks the mantle of substance once all cosmetic hype, bells, and whistles are removed.

Slick diagrams, catchy slogans and colorful imagination can't guarantee scientific respectability. Against impossible odds, evolution attracts minds to the obscenity that an ancestral, grand pappy fish spawned descendant humanity by chance.

Truth stands tall in three dimensions; it needs no defense.

Falsehood collapses on itself, melting to nothingness when exposed to the scrutiny of hard evidence under enlightening rays of the noonday sun.

Physician Sean Pitman articulates the reality that the fossil record doesn't rank with Mendel's Law of Genetics in evaluating evolution.

"…Interpretations of fossils and the geologic column is not as definitively precise and conclusive as dealing with genetics and the Darwinian mechanism…" [26]

Pitman went on to charge, *"If Darwinian-style evolution happens or doesn't happen, it happens or doesn't happen genetically…*

"The Darwinian mechanism of random mutations combined with natural selection was statistically untenable — dramatically so. Given billions or even trillions of years of time, it was hopelessly inadequate to explain the origin of novel functional biological information beyond very very low levels of functional complexity…

"I was especially shocked at the use, by modern scientists, of Mendelian variation as a basis for Darwinian-style evolution over time. Mendelian variation isn't evolution at all. It is simply a difference in expression of the same underlying gene pool of options where the gene pool itself doesn't change." [26]

Louis Bounoure's blunt analysis concluded, evolution *"is a fairy tale for grown-ups. This theory has helped nothing in the progress of science. It is useless."* [27]

The late Stephen J. Gould, articulate Harvard biologist, undercut the core essence of evolution's *synthetic theory* with cold-eyed logic.

"You don't make new species by mutating the species…A mutation is not the cause of evolutionary change." [28]

So much for evolution's iconic lynchpin!

X

The Laughing Mouse

Irreducible Complexity

*"…Natural selection can only choose systems that are already working…
if a biological system cannot be produced gradually
it would have to arise as an integrated unit, in one fell swoop,
for natural selection to have anything to act on."* [1]

Michael J. Behe

© Jemini Joseph

The Paris Eiffel Tower, icon of *"irreducible complexity!"*

Irreducibly complex mechanisms, living or man-made, require a system of inter-dependent components in order to function.

Today's world turns on personal computers. Take away the hard drive and the irreducibly complex machine would become useless junk. Even without a keyboard or the electric mouse and it would be severely impaired.

Like a man-made computer, animal and plant life forms could not exist without all DNA components in place. Mutations spell disaster!

A mousetrap is as simple as any human designed mechanism. When all its components function in unison, the trap can be deadly to mice. Remove any single part, and the trap morphs into a harmless pile of pieces.

Biochemist Michael J. Behe, introduced landmark phraseology when he suggested molecular systems in living organisms exist *"irreducibly complex."* He illustrates the world of molecular machines by citing the man-made mousetrap as a prime example.

One skeptic suggested the trap could do without the wooden base by stapling the trap's metal parts to the floor. The argument fails---the floor itself becomes an essential part to the irreducibly complex mechanism.

The simple but *"irreducibly complex"* mousetrap consists of five basic parts anchored by four staples. It requires the complete, designed package of parts for minimal function. A spring too weak, a trigger too short, or a missing staple and the mouse would steal the cheese, survive unscathed and scoot away safely with the last laugh.

A fully functioning mousetrap is the product of human intelligence. No one argues that inanimate machines design and build themselves without input from an intelligent source. Design and construction of molecular mechanisms also requires intelligent input.

Neither mousetraps nor computers can't design themselves without input from an intelligent source. Living mechanisms display designs infinitely more complex than objects designed by human intelligence. Assembling operational living systems requires insight beyond nature's accidental whims.

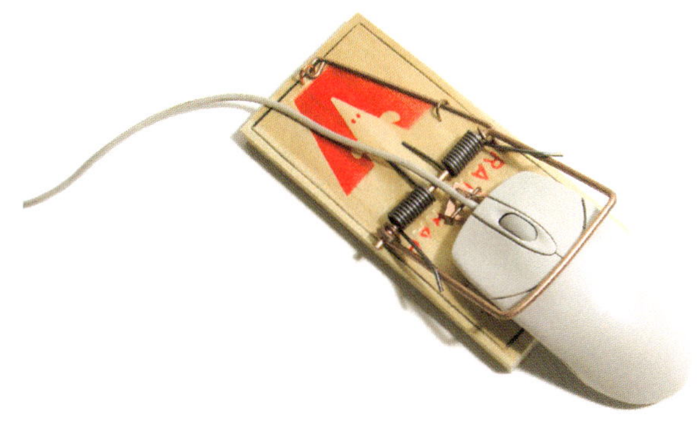

**Remove any part of an *"irreducibly complex"* mousetrap
and the mouse walks away with the cheese and the last laugh.**

"Canyons separating everyday life forms have their counterparts in the canyons that separate biological systems on a microscopic scale. Like a fractal pattern in mathematics …unbridgeable chasms occur even at the tiniest level of life." [2]

Part-way-there transitionals couldn't survive the gradual journey whether whales transiting from bears or birds from dinosaurs!

**A prime example of *irreducible complexity* in the living world
is the bacterial flagellum, a molecular machine which functions
with 40 separate components, 30 of which are unique.**

The flagellum's microscopic propeller cranks at an amazing 100,000 RPM's per minute and is capable of stopping and reversing on a quarter turn.

If any single component is missing, the entire organism breaks down. Time is of the essence. Its an all or nothing-at-all scenario; each component must be in place simultaneously for the living mechanism to function. [3]

This fact-of-life flies in the face of Darwin's postulate alleging complex organs could be *"formed by numerous, successive, slight modifications."*

"Irreducible complexity" is the joker in evolution's card deck! Organs of living systems cannot evade this cardinal principal. Human eyes could not see without optic nerves and could not survive without being bathed by soothing liquid flowing from lachrymal glands.

Behe cites the chemical apparatus marshaled by the half-inch bombardier beetle as an irreducibly complex system in action. When threatened by an enemy it can release a scalding hot liquid as a defense.

"The components of the system are (1) hydrogen peroxide and hydroquinone, which are produced by the secretory lobes; (2) the enzyme catalysts, which are made by the ectodermal glands; (3) the collecting vesicle; (4) the sphincter muscle; (5) the explosion chamber; and (6) the outlet duct." [4]

Behe inquires, *"What exactly are the stages of beetle evolution, in all their complex glory? Second, given these stages, how does Darwinism get us from one to the next?"* [4]

"As the number of required parts increases, the difficulty of gradually putting the system together skyrockets, and the likelihood of indirect scenarios plummets.

"Darwin looks more and more forlorn." [5]

Irreducibly complex blood powers animal life.

Custom-designed formulas match the needs of each species.

One-of-four blood types flow through human arteries and veins. Egyptian mummies share identical bloodlines with modern *Homo sapiens*---unchanged after thousands of years. Transfusing blood from ape to man can kill; extreme blood loss can kill; blood too thin to clot can kill; and blood that clots at the wrong time and place can kill.

Behe emphasizes the magical qualities of clotting.

"If blood congeals at the wrong time or place…the clot may block circulation as it does in heart attacks and strokes…A clot has to stop bleeding all along the length of the cut, sealing it completely. Yet blood clotting must be confined to the cut or the entire blood system of the animal might solidify, killing it." [6]

Properly functioning blood clotting in the human circulatory system that can save life involves no fewer than twenty steps.

"The blood coagulation cascade" utilizes *"proteins in promoting clot formation"* and proteins *"involved in the prevention, localization, or removal of blood clots."* [7]

"None of the cascade proteins are used for anything but controlling the formation of a blood clot. Yet in the absence of any one of the components, blood does not clot, and the system fails…Not only is the entire blood-clotting system irreducibly complex, but so is each step in the pathway." [8]

"Remember a mousetrap spring might in some way resemble a clock spring, and a crowbar might resemble a mousetrap hammer, but the similarities say nothing about how a mousetrap is produced.

"In order to claim that a species developed gradually by a Darwinian mechanism a person must show that the function of the system could 'have been formed by numerous successive, slight modifications.' " [9]

Mathematical odds say, "no way!"

Blood clotting could not evolve *"by numerous successive, slight modifications."*

Tissue Plasminogen Activator binds to several substances including fibrin. TPA *"has four different types of domains…the odds of getting those four domains together is 30,000 to the fourth power.*

"Now if the Irish Sweepstakes had odds of winning of one-tenth to the eighteenth power, and if a million people played the lottery each year, it would take an average of about a thousand billion years before anyone (not just a particular person) won the lottery.

"A thousand billion years is roughly a hundred times the current estimate of the age of the universe." [10]

"We calculated the odds of getting TPA alone to be one-tenth to the eighteenth power; the odds of getting TPA and its activator together would be about one-tenth to the thirty-sixth power…

"Such an event would not be expected to happen even if the universe's ten-billion year life compressed into a single second and relived for every second for ten billion years…The fact is, no one on earth has the vaguest idea how the cascade coagulation came to be." [11]

Michael Denton, internationally recognized authority respecting molecular machinery, dismisses *"gradual accumulation of random changes"* as unsustainable in light of irreducible complexity.

"My fundamental problem with the theory is that there are so many highly complicated organs, systems and structures, from the nature of the lung of a bird, to the eye of the rock lobster, for which I cannot conceive of how these things have come about in terms of a gradual accumulation of random changes.

"It strikes me as being a flagrant denial of common sense to swallow that all these things were built up by accumulative small random changes. This is simply a nonsensical claim…a huge number of highly complex systems in nature cannot be plausibly accounted for in terms of a gradual build-up of small random mutations…

"Everybody knows the lung of the bird is unique in being a circulatory lung rather than a bellows lung…It doesn't require a great deal of profound knowledge of biology to see that an organ, which is so central to the physiology of any higher organism, its drastic modification in that way by a series of small events is almost inconceivable.

"This is something we can't throw under the carpet again because, basically, as Darwin said, if any organ can be shown to be incapable of being achieved gradually in little steps, his theory would be totally overthrown." [12]

The *"break down"* time that concerned Darwin, knocks at the door!

Michael Denton's analysis takes Darwin at his word, challenging gradualism's imaginary make-believe as a substitute for fact.

"If anyone was chasing a phantom or retreating from empiricism it was surely Darwin, who himself freely admitted that he had absolutely no hard empirical evidence that any of the major evolutionary transformations he proposed had ever actually occurred.

It was Darwin himself, the evolutionist, who admitted in a letter to Asa Gray, that one's *'imagination must fill up the very wide blanks.'* " [13]

Donald R. Moeller, a science researcher blessed with professional credentials as both physician and dentist, shoots down any possibility that unique dentition sizes and shapes result from mutations.

Moeller's logic devastates neo-Darwinism's core dogma's reliance on
"numerous, successive, slight modifications!" Modern craniofacial/maxillafacial
genetics demonstrates that pleiotropy is present in approximately one
hundred genetic disorders.

*"Thus the simplistic idea of genetic mutations being able to cause only incremental
small useful changes in the occlusion and/or jaw relationships is not supported by current
research. There are no known mutations affecting single tooth morphology or single tooth
enamel microstructure.*

*"Multiple examples of malposed teeth, cysts in jaws, retained deciduous teeth,
maxillary-mandibular growth and size discoordination, losses of entire classes of teeth,
variation in eruption height of the various classes of teeth, tooth size arch-discrepancy
variation, animal size, and tooth size coordination, should be in the fossil record.*

"This evidence is not seen...

*"There is also no known genetic or developmental process to suggest a legitimate
mechanism to support an evolutionary basis for the development of the precision exhibited
by the dental apparatus."* [14]

Darwin assembled a wish list of uncorroborated assertions, predictions,
preconceptions, and assumptions in weaving together the whole cloth of
accidental life without design or designer.

He *"ransacked other spheres of practical research work for ideas…*

*"But his whole resulting scheme remains, to this day, foreign to scientifically
established zoology, since actual changes of species by such means are still unknown."* [15]

All living mechanisms function irreducibly complex. Molecular
mechanism design requires intelligence beyond the quirks of nature's
whims. Assembling complex systems requires oversight.

It's all or nothing at all with the human eye. Sight is contingent upon
every lens, blood vessel, and nerve in place and fully operational.

The same with ears, hearts, lungs and teeth.

There is no reason for an infinitely wise and all-powerful Creator to require millions of years to assemble a brain or any life system.

Recognition of Supreme power accepts the plausibility of a creation event at the command of the Lord God Almighty in the blink of an eye. Anything less suggests the eye is blind to the power.

© Ruth Page Johns

**Since completion in 1937,
San Francisco's graceful Golden Gate Bridge
depends on every thread of its steel cables for survival.**

The eye of an arthropod also astounds.

Arthropod eyes *"have always been complex---and there have always been arthropods…The bee's ability to convey the location of a food source to fellow workers via a sophisticated 'dance' is legendary."* [16]

Trilobites *"could see an undistorted image under water…with undistorted vision in all directions…to determine distance…while at the same time have the optimum sensor for motion detection."* [17]

Darwin's gradual eye evolution theory, beginning with tissue theoretically able to recognize light, couldn't do the trick for an owl or a hawk.

Park Rangers at Tennessee's Reelfoot Lake protect wildlife and restore the injured. When recovery is complete, healthy eagles are released to circle the heavens.

A screech owl, surviving a collision with a car, lost an eye in the accident. Park Rangers explain that a bird (like the red-tailed hawk, blessed with acute vision extending a mile in any direction) couldn't survive in the wild with a single eye.

A hawk, flapping half-formed wings while trying to hunt with eyes scarcely able to detect night from day, couldn't feed or protect itself. A less "evolved" parent, even if miraculously alive, couldn't help its crippled offspring.

Whether an eye, wing, or foot, however long or short its "evolution" time, it could never function in part-way-complete transitional format. Extinction intrudes, long before expiration of evolution's postulated mega-million years of deep time.

Irreducible complexity comes into play ecologically.

Swallows swoop to mud puddles, scavenging nest material. Eagles build 3-foot wide nests in treetops. Parrots arrayed in multi-colored hues and oversize beaks mimic human sounds.

Virtually nothing, either in the fossil record, molecular biology, or today's experience, confirms existence of partially or non-functioning intermediates linking these magnificent birds to reptile ancestors.

© Elena Eliseeva

Life did not require vast chunks of deep time to originate abruptly, fully-formed and in perfect balance. Multiple plant and animal kinds, appearing simultaneously and functioning in concert, provided an *irreducibly complex* eco-system from the beginning.

Gradualism is out the window if tested against this all-or-nothing measure. A chaotic composite of non-functioning organic systems would throw ecological balance out-of-kilter.

Peacocks spread their array of iridescent feathers while notifying neighborhoods of intruders with ear-piercing screams, testifying to their irreducibly complex status.

Natural selection works contrary to gradualism by rejecting a non-functioning component part, any incomplete, irreducibly complex cell, organ, or system would be discarded by the natural selection process.

130

Prolonged failure to function is a recipe for the destruction of most any living system trapped interminably in a prolonged series of intermediate states. A creature attempting to evolve a wing from a leg, left unable to walk or to fly, is destined to die. When the delicate balance of a complex life is disrupted by dysfunction, life terminates.

Conversely, life, witjh its sophisticated systems (eg, the arthropod's vision), can't spring from an incomplete organism missing critical parts.

Nature bursts with dynamic change, but irreducible complexity prevents activation of the very natural selection critical to evolution. Half-formed hearts and lungs are no more practical than half-built automobiles.

Thanks to genes present in every genome from the beginning, life kinds adjust, assuring survival while maximizing nature's smorgasbord of colors, shapes and sizes. Adjustment capability doesn't result from mutations but comes from the genetic code already inherent in every genome.

Darwin contemporary, Sir William Dawson, warned that when evolution was scrutinized, *the whole fabric would melt away like a vision.*

"This evolutionist doctrine is itself one of the strangest phenomena of humanity…a system destitute of any shadow of proof, and supported merely by vague analogies and figures of speech…now no one pretends that they rest on facts actually observed…

"…Evolution, as an hypothesis, has no basis in experience or in scientific fact, and that its imagined series of transmutations has breaks which cannot be filled." [18]

Irreducible complexity underscores an all-or-nothing reality!

Sir William's candid assessment proved to be on the money. The century since Darwin ignored *irreducible complexity* in his eagerness to cast his vote for molecule-to-man by accident has not been kind to the theory.

Biologist, Søren Løvtrup, scoffed at evolution's unproven scenario.

"I believe that one day the Darwinian myth will be ranked the greatest deceit in the history of science." [19]

Recent scientific setbacks gnawing away at evolution's core fabric were spotlighted by an "Intelligent Design" advocate, writing in 2009.

"In a year in which Darwin's disciples were celebrating the 200th anniversary of his birth that the 150th anniversary of the publication of On the Origin of Species, mainstream scientific journals published articles declaring:

(1) the modern synthesis was dead,(2) Darwin's tree of life should be abandoned, (3) new "missing links" were a bust, (4) limits to Darwinism were demonstrated in the lab, (5) evolutionary icons like the peppered moths reverted back to their old colors, (6) the Cambrian Explosion lacks any plausible materialist explanation, and (7) an interdisciplinary revolution is occurring in biology that rejects the reductionist paradigm of Darwinian evolution." [20]

---Darwin drew a blank trying to explain the origin of first life on earth.

---Evolutionist Ernst Mayr concedes, *"All species are separated from each other by bridgeless gaps; intermediates between species are not observed."* [21]

---Millions of complex life kinds appear in the fossil record abruptly, fully-formed, and without a trace of fossil ancestry.

---Contrary to Darwin's prediction, species, closely resembling fossil ancestors, continue thriving, very much alive.

---Darwin's fictional chain of organic life, shows a fossil cupboard virtually bare Gaps dominate. Discontinuity reigns.

---Molecular biology's electron microscope uncovered the magnificent complexity of the living cell with DNA's language of life, demolishing the primitive notion of the supposedly "simple" cell.

---The Modern Synthetic Theory of Evolution dug deep attempting to salvage Darwinian thought by coupling *mutations* with *natural selection* as twin forces driving evolutionary change. This one-step-forward "solution" exposes a trail of two-step-back problems.

(1) Mutations typically corrupt genetic information, never adding new information or upgrading the quality of the organism.

(2) Five-hundred-million years is not enough time to accumulate the millions of allegedly "beneficial" mutations essential for random chance to accidentally evolve an *irreducibly complex* human from a fish.

The Modern Synthetic Theory of Evolution deserves last rites!

"...According To Its Kind" [1]

Stasis

"We may safely infer that not one living species will transmit its unaltered likeness to a distant futurity ...." [2]

Charles Darwin

© Warren L. Johns

The Pre-Cambrian landscape offers a *"paleontological desert,"* virtually barren of identifiable fossil ancestors. Rather than a reliable geologic column brimming with obvious transitionals, stasis intrudes.

"Almost abrupt appearance of the major animal groups," from as many as 50 phyla, ranks as *"one of the most difficult problems in evolutionary paleontology."* [3]

Not only is this *"not an obviously predicable outcome of evolution,"* [4] descendents of many Cambrian era species live today, virtually unchanged from their fossil ancestors.

Today's world is loaded with descendants of Cambrian animal and plant species, relatively unchanged from their fossil ancestors, devastating an already shaky theory predicated on a persistent chain of transition.

Unlike the theory's prediction that ancestor species would be swept away genetically and be replaced by descendant generations of new and improved life kinds, earth's strata shouts "stasis."

Discontinuity and stasis trump gradualistic evolution!

The jaundiced eye of Henry Gee viewed attempts to *"take a line of fossils and claim that they represent a lineage"* as a less than scientific *"assertion that carries the same validity as a bedtime story."* [5]

Gee offers persuasive evidence!

"All fossil roundworms found to date are indistinguishable from modern forms." Fossils conventionally dated *"more than 3 billion years old are virtually identical to organisms, called cyanobacteria, living today."* [6]

The abrupt and relatively simultaneous appearance of several thousand different plant and animal species, without evidence of transition ancestry prior to the Cambrian, dealt a severe blow to the heart of evolution theory.

In addition to the once-considered extinct coelacanth, fossil cemeteries include crustaceans, millipedes, horseshoe crabs, snails, bivalves, squid-like belemnolds, nautilus-like ammonolds and mollusks.

Moving up the "geologic column" from the deepest sedimentary layers, stasis abounds: stromatolites, brittlestars, sea lilies, dragonflies, sycamore tree leaves, wasps, and sand dollars.

**Dragonflies never received the message to evolve.
Today's descendants replicate their ancient fossil ancestors.**

Multi-celled animal embryos, no bigger than a grain of sand, have been discovered in China, preserved in calcium phosphate and dated at the edge of the Cambrian/Precambrian time frame—marking the prolific explosion of organic life during which *"virtually all the major animal body plans seen on Earth today blossomed in a sudden riotous evolutionary springtime."*[7]

Shixue Hu of China's Chengdu Geological Center reports discovering a cemetery of 20,000 fossils, many fully intact, buried fifty feet deep in a Luoping mountain in southwestern China.

Duly awarded a 250 million year age, the treasure trove of antiquity includes stunning evidence at odds with Darwin's unproven prediction that *"not one living species will transmit its unaltered likeness to a distant futurity."* [8]

There they were, allegedly 250 millions after the fact: clams, oysters, snails, and even a duplicate of today's coelacanth fish, virtual mirror images of today's life forms. Stasis prevails over those missing transitionals.

The find's stunner was the *"soft tissues"* found on some of the fossils! [9]

"Soft tissues" that escaped decay for 250 million years? Really?

Beyond resolving the *"far higher problem of the essence or origin of life"* the *"sudden"* appearance *"of the higher plants"* mystified the naturalist. *"Nothing is more extraordinary in the history of the Vegetable Kingdom …than the apparently very sudden or abrupt development of the higher plants."* [10]

In an 1879 letter to Joseph Hooker, a troubled Darwin expressed perplexity that flowering plant ancestry seemed missing from the fossil record. *"The rapid development, as far as we can judge, of all the higher plants within recent geological times is an abominable mystery."* [11]

The *"abominable mystery"* is apparent in the glaring omission from Darwin's *"tree of life"* of any rational explanation for the origin of the entire Plant Kingdom.

Wollemi pine trees, conventionally dated at 150 million years before the present, were long considered extinct. To everyone's surprise, hardy Wollemi pines have been discovered alive and well.[12] Thriving in an obscure Australian grove, the Wollemi remain virtually unchanged for the entire world to see and admire.

That scraggly Bristlecone pine, a stalwart of the Plant Kingdom, thrives high in the Sierras, with rings suggesting an age reaching back 4,000 years before the present.

And where did that Giant sequoia, towering a football field length into the sky, inherit its 3,500 year-old roots?

The oldest known life forms living on earth belong to the Plant Kingdom.

Nothing has evolved from Wollemi pines, Bristlecone pines, or Giant Sequoias in the last 4,000 years, so where does Darwin predict evidence of the radical change?

As to flowers such as crocus, roses, orchids, and the purple blossoms that spring from wisteria vines, what is the common ancestor?

Then there are those food-producing plants!

Potatoes that grow in the ground; walnuts, apples, peaches and apples from trees; grains like wheat and corn; and grapes and raspberries from vines. Darwin didn't find the time to identify his *"abominable mystery"* with the common ancestry notion!

Evolution theory has done little to explain the Plant Kingdom and just where it might have diverged from the Animal Kingdom after that mysterious first living cell allegedly created itself accidentally in the imaginary *"warm little pond."*

Perhaps its easier and more news worthy for evolutionists to focus on fish, dinosaurs, apes and monkeys. Whatever the reason, the *"abominable mystery"* remains an unsolved *"mystery."*

Like the world of animals, the plant kingdom offers no refuge to an inadequate, if not unsustainable premise. Late in the 20th century, a Cambridge University Botany scientist ventured, *"…to the unprejudiced, the fossil record of plants favours [sic] special creation."* [13]

Discontinuity rather than a continuous chain of evolving organic life forms dominate fossil records of both plant and animal kingdoms.

No question, many plant and animal species have suffered extinction.

At the same time, close matches with descendant organisms populating today's world, emerge from ancient strata. A cast of thousands thrives profusely, perpetuating virtual resemblance to ancient ancestors.

Recognizing extinction while either ignoring, or unaware of the pervasiveness of stasis, Darwin overreached, proclaiming without equivocation, *"not one living species"* would escape to the future.

© **Chuck Nelson**

Evidence preserved in stone contradicts Darwin's guess.

The fossil record didn't oblige his expectations!

Extinctions? Of course!

Failure to transmit an *"unaltered likeness to a distant futurity?"*

Gross error!

Instead of a reliable geologic column brimming with obvious transitionals as mandated by evolution theory, stasis intrudes. The fossil record bulges with evidence of "stasis" while fossil chains demonstrating organic continuity bridging across-the-board, molecule-to-man transitions remain virtually invisible.

Conjectured chunks of deep time made anything seemingly possible--- even exaggerated leaps over vast chasms of biological diversity. Darwin predicted, given enough time, every parent species would be wiped out by superior descendants.

Sidestepping stasis reality, Darwom built his theme on the uncorroborated inference that *"the production of new forms has caused the extinction of about the same number of old forms."* [14]

Contrary to this fantasy, millions of fossils shout stasis!

After 425 million years of conventional time, the Ostracode, remains unchanged. [15]

Ancestral alligators, oysters, sea urchins, horseshoe crabs, bowfins, Australian lungfish, sturgeons, crinoids, bats, arrow worms, opossums, starfish, corals, and the platypus weave a consistent pattern of resemblance with descendants.

Hopelessly out-of-kilter with Darwin's prediction *"…that not one living species will transmit its unaltered likeness to a distant futurity,"* [2] an allegedly 360 million-year-old fossil shrimp, discovered in Oklahoma, appears to be a look-alike replica of shrimp living today on ocean seafloors. [16]

Devonian era fossils, virtually identical to animal species living in today's world, not only demonstrate evidence of stasis but also can be construed to call into question evolution's conventional theory that amphibians descended from fish ancestors. [17]

Evolutionist Henry Gee's ranked the horseshoe crab as yet another species that dodged extinction's bullet, *"…An animal that has not changed its basic form for hundreds of millions of years."* [18]

Steven Austin, devout creationist with impeccable science credentials, confirms that *"Clams have always been clams; brachiopods have always been brachiopods; fish have always been fish."* [19]

Even the lowly cockroach survives unscathed from mega-time's ravage

The Western Pacific's nautilus is a dead ringer for its long-fossilized ancestor. *"In every way they are virtually identical to the living chambered nautilus. The creature that swims in our oceans today is the same one that was swimming 100 million years ago."* [20]

Fossil insect species identified in ancient Scandinavian amber bear a striking resemblance to today's descendants.

"Lungfish almost identical to those of modern Africa are found as fossils in the rocks of the Devonian era…alongside fossils of the earliest amphibians and the very fish groups from which the amphibia supposedly arose." [21]

Greenling damselflies, tiny insects with 22mm wingspans, supposedly went extinct 250 million years ago, but there they are, still flying in Australia. [22]

While Darwin imagined extinctions, 1840's era excavators for England's Great Western Railway stumbled across a fossil cemetery *"where thousands of Jurassic fossils with preserved soft tissues were found"* and then lost until rediscovered by a team led by Phil Wilby.

"…Literally millions of these animals were dying out or being killed in this precise area and we don't know what that is." [23]

© Rob Hainer

This dragonfly seems unaware he replicates ancient ancestors.

Not only did the ancient squid resemble its modern counterpart, but also its inch-long ink sac delivered the formula for the ink used by an artist to sketch its design and to sign its scientific name.

Wilby marveled!

"It is difficult to imagine how you can have something as soft and sloppy as an ink sac fossilized in three dimension, still black, and inside a rock that is 150 million years old. The structure is similar to ink from a modern squid so we can write with it." [23]

As with non-existent "organic chains" in the animal kingdom in the fossil record, the plant kingdom shows comparable shortfall. Quite the contrary, stability reins supreme.

Hickory, walnut, magnolia, and gingko trees, along with grape vines, and water lilies boast ancient fossil ancestries essentially matching twenty-first century's distinctive counterparts.

Remnants of trees, embedded deep in coal seams since ancient times, include a diverse roster of plant kingdom favorites: sassafras, laurel, poplar, willow, maple, beech, birch and elm. Descendants of these familiar names, survive virtually unchanged in today's forests, oblivious to Darwin's prediction of their certain demise.

The paucity of fossil evidence corroborating his theory troubled Darwin. He recognized fossils found during his lifetime revealed few if any *"finely-graduated organic chain."* He set the bar high, putting all his cards on the table with one giant leap of faith!

"If my theory be true, numberless intermediate varieties, linking closely together all the species of the same group, must assuredly have existed" [24]

A frustrated Darwin bemoaned the glaring gaps between theory and fact. *"Geological research…does not yield the infinitely many fine gradations between past and present species required on the theory…Why do we not find beneath this system great piles of strata stored with the remains of the progenitors of the Cambrian fossils?"* [25]

The shortfall nagged at the naturalist's mind, leading him to admit he recognized the missing chain of transitional links was *"the most obvious and serious objection which can be urged against the theory."* [26]

Since the publication of *Origin*, millions of fossils have been rescued from rocky hiding places. But the missing organic chains of transitional links continue AWOL.

Beyond defining *"the most obvious and serious objection which can be urged against the theory"* he acknowledged *"If it could demonstrated that any complex organ existed, which could not possibly have been formed by numerous, successive, slight modification, my theory would absolutely break down."* [27]

To validate evolution theory, all life forms would be transiting by gradual increments of *"successive, slight modification,"* and the fossil record would be brimming with intermediate life forms.

Instead, the fossil cupboard overflows with extinctions and stasis pointing to discontinuity rather than the expected *"inconceivably great"* number of *"transitional links."*

Henry Gee recognizes chasms of discontinuity acknowledging, *"…it is impossible to know for certain whether one species is the ancestor of another."* [28]

A realist, he points out *"…that adaptive scenarios are simply justifications for particular arrangements of fossils made after the fact, and which rely for their justification on authority rather than on testable hypotheses."* [29]

Ernst Mayr concluded, *"All species are separated from each other by bridgeless gaps; intermediates between species are not observed."* [30]

Unwilling to discard the toil of a lifetime, Darwin backtracked suggesting nineteenth century fossil discoveries must be woefully incomplete. He counted on future discoveries of fossil evidence to salvage his theory. Fortified with a visionary's optimism, Darwin shrugged off the shortfall, ascribing the missing branches in his conjectured fossil "tree of life" to the *"extreme imperfection of the geological record."* [31]

He ignored obvious reality: that *"inconceivably great"* number of *"transitional links"* never existed! But no amount of wishful thinking could mask imagination's virtual blank screen.

Early in century twenty-one, with more than two billion fossils exposed to the light of day, those elusive *"transitional links"* continue missing in action.

"Surprising as it may seem, the only real evidence for the geological succession of life, as represented by the timetable, is found in the mind of the geologist and on the paper upon which the chart is drawn. Nowhere in the earth is the complete succession of fossils found as they are portrayed in the chart." [32]

Scraping the bottom of the bone barrel, efforts continue fruitless in the quest for supposed missing links. Rather than a roster of conclusive evidence, evolution loyalists have had to settle for the rare and the dubious.

Using fossil bone scraps in an attempt to link birds to dinosaurs and whales to some land-based ancestor with a foot is a far cry from reliable evidence confirming *"numberless …intermediate"* transitionals.

The idea that some mammal with a foot evolved into whale status affronts the Genesis account that *"…God created the great creatures of the sea and every living and moving thing with which the water teems, according to their kinds. and every winged bird according to its kind.."* [33]

Darwin's imagination demeans and disrespects human life while fostering irreverent contempt for the Creator!

As recently as 1982, *"The fossils that decorate our family tree are so scarce that there are still more scientists than specimens."* [34]

Since the living world is hardly awash with definitive examples of evolution in action, it remains for the fossil record to deliver more than unsubstantiated conjecture.

Whether exposed to the shovels digging for clues in fossil fields or to spotlight glare of microscopes rolling back the mysteries of molecular biology, Darwin's elaborate predictions have not fared well.

Absent requisite transitionals, Darwin's dream dangles in limbo, clinging to an unraveling thread. Continuous chains of organic life forms have yet to be forged from unrelated fragments of fossil bones.

Borrowing a quip from Mark Twain's humor, evolution's supposed mantle of authenticity is *"greatly exaggerated."*

Or in the phraseology of gritty Texas lingo, evolution's unproved assumptions qualify as *"all hat and no cattle."*

Endnotes

Chapter I

1---Geroger Frederic Handel, *The Messiah* (1741-42).
2---*The Holy Bible, King James Version,* John 14:6.
3---*The Holy Bible,* New International Version, John 18:38.
4---*The Holy Bible,* King James Version, Mathew 5:3-10
4---*The Holy* Bible, New International Version, Mathew 5:14 & 15.

Chapter II

1---Søren Løvtrup (Swedish biologist), *Darwinism: The Refutation of a Myth*(New York: Croom Helm, 1987), p. 422.
2---Charles Darwin, *On the Origin of Species by Means of Natural Selection, or the Preservation of Favored Races in the Struggle for Life*, (First Edition facsimile, 1859, Cambridge: Harvard University Press) 184.
3---*The Holy Bible*, King James Version. Genesis 1:21.
4---Charles Darwin, *Origin of Species,* 647.
5---__________, *Descent of Man,* Vol. I, 169.
6---__________, *Descent,* Vol. I, 168.
7---__________, *Descent,* Vol. 1, 216.
8---__________, *Descent,* Vol. I, 178.
8---__________, *Descent,* Vol. I, 201.
10---__________, *Descent,* Vol. II, 328.
11---__________, Descent, Vol. II, 327, 328.
13---David Quammen, *The Reluctant Mr. Darwin* (New York: Atlas Books, W.W. Norton & Company, 2006) 28.
14---Charles Darwin, *Origin,* 247.
15---__________, *Origin,* 453.
16--Bert Thompson, *The Scientific Case for Creation* (Montgomery, Alabama: Apologetic Press, Inc., 2002) 124.
17---Noble, et. al, *Parasitology,* sixth edition, "Evolution of Parasitism," Lea and Febiger, 1989, 516, as cited by Frank Sherwin, *Origins Issues,* "Natural Selection's Role in the Real World."
18---Colin Patterson, "Cladistics," *The Listener* (1982) 106:390.
19---Charles Darwin, Letter to J.D. Hooker [1 February] 1871, in Darwin, F., ed., *The Life and Letters of Charles Darwin,* [1898], (New York: Basic Books, Vol. II, 19590, reprint, 202-203.
20---Charles Darwin, *Descent*, Vol. II, 389.
21---__________, *Descent,* Vol. 1, 206.
22---__________, *Descent,* Vol. 1, 213.
23---__________, *Origin*, 649.
24---Michael Denton, *Evolution: a Theory in Crisis*, (Bethesda, Maryland: Adler & Adler, 1986) 117, citing Darwin, C., (1858) in a letter to Asa Gray, 5 September, 1857, *Zoologist*, 16:1697-99, see 1699.
25---Charles Darwin to Asa Gray, cited by Adrian Desmond and James Moore, *Darwin,* (New York: W.W. Norton and Company, 1991) 456.
26---Desmond & Moore, *Darwin, 475.*
27---__________, *Darwin,* 477.
28---Sharon Begley, "How to Think Like a Scientist," *Newsweek,* July 9, 2007.

Chapter III

1---This thought is attributed to Mark Cahill, but the precise reference not known.
2---Martin Rees and Priyamvada Natarajan, "Invisible Universe," *Discover,* (December, 2003) 18.
3---Charles Siebert, "Unintelligent Design," *Discover*, Vol. 27, No. 3, March, 2006, 31, 34.
4---Brian Greene, Professor of mathematics and physics, Columbia University, as quoted by Carl Warner, *Living Fossils, Evolution: the Grand Experiment,* (Green Forest, AR: New Leaf Press, 2008) 4.

5---Peter Coles, "Boomtime," *New Scientist,* March 3, 2007.

6---Dean L. Overman, 59, citing Hoyle and Wickramasinge, 148, 24, 150, 30, and 31 as quoted in Thaxton, Bradley, and Olsen, 196.

7---Harold J. Morowitz, *Energy Flow in Biology* (New York: Academic Press, 1968); cited by Coffin, 376.

8---John Keosian, In Haruhiko Nada, ed., *Origin of Life* (Tokyo: Center for Academic Publications, Japan Scientific Publications Press, 1978) 573, 574, quoted by Coffin, 377.

9---Walter L. Bradley and Charles B. Thaxton, "Information and the Origin of Life," in The Creation Hypothesis, ed. J.P. Moreland (Downers Grove, Illinois: Inter Varsity Press, 1994) 190 as quoted by Overman, 62.

10---Dean L. Overman, *A Case Against Accident and Self-Organization,* 58, 59.

11---Charles Darwin, *Origin*, 649.

12---__________, Letter to J.D. Hooker [1 February] 1871, in Darwin, F., ed., *The Life and Letters of Charles Darwin,* [1898], (New York: Basic Books, Vol. II, 19590, 202-203.

13---Dean L. Overman, *A Case Against Accident and Self-Organization* (New York: Rowman & Littlefield Publishers, Inc., 1997) 38. See Aleksander I. Oparin, *The Origin of Life* (New York: Dover Publications, Inc., 1938).

14---Charles B. Thaxton, Walter L. Bradley and Roger L. Olsen, *The Mystery of Life's Origin* (New York: Philosophical Library, 1984) 182, 183, 185, quoted by Bert Thompson, 80.30—Denton, 261.

15---Michael Denton, *Evolution, A Theory in Crisis,* 261.

16---Hubert P. Yockey, *Information Theory and Molecular Biology* (Cambridge: Cambridge University Press, 1992), 257, quoted by Overman, 61, 62.

17---Gunter Wachtershauser, Letter to Editor, *Science*, 25 October 2002, vol. 298.

18---Larry A. Witham, *Where Darwin Meets the Bible* (New York: Oxford University Press, 2002) 129.

19---Sir William Dawson, *The Story of Earth and Man* (New York: Harper and Brothers, 1887) 317, 322, 330, 339.

20---Paul Davies, *The Cosmic Blueprint: New Discoveries in Nature's Ability to Order the Universe* (New York: Simon & Schuster, 1988) 203 and *The Mind of God* (New York: Simon & Schuster, 1992) 232.

21---Michael J. Murray, *Reason for the Hope Within* (Grand Rapids, Michigan: Eerdmans, 1999) 61-62

22---__________, *Reason for the Hope Within,* 48.

23---Dean L. Overman, *A Case Against Accident and Self-Organization*, 40, 41.

24---Percival Davis, Dean H. Kenyon, and Charles B. Thaxton, Academic Editor, *Of Pandas and People* (Dallas: Haughton Publishing Company, 1993) 3, 4.

25---Ralph O. Muncaster, *Creation Versus Evolution* (Mission Viejo, Calif.: Strong Basis to Believe, 1997) 17.

26---Percival Davis, Dean H. Kenyon, and Charles B. Thaxton, Academic Editor, *Of Pandas and People,* 3.

27---Dean L. Overman, *A Case Against Accident and Self-Organization,* 42.

28---Michael Denton, *Evolution: A Theory in Crisis* (Bethesda, Md.: Adler & Adler,1986) 261, 262.

29---*The Holy Bible*, *New International Version* (Grand Rapids: Zondervan Bible Publishers, 1983), Genesis 1:2.

30---Rick Weiss, "Water Scarcity Prompts Scientists to Look Down," *Washington Post*, March 10, 2003, A-11.

31---Robert Jastrow, *Until the Sun Dies* (New York: W.W. Norton, 1977) 62, 63 as quoted by Bert Thompson, *The Scientific Case for Creation*, 76.

32---Chandra Wickramasinghe "Threats on Life of Controversial Astronomer," *New Scientist,* January 21, 1982, 140, as quoted by Overman, 60.

Chapter IV

1---Michael Denton, *Evolution: A Theory in Crisis*, (Bethesda, Md.: Adler & Adler, 1986) 96, 323.

2---Charles Robert Darwin, *Descent of Man,* Vol. II, 327.

3---Bert Thompson & Brad Harrub, "15 Answers to John Rennie and *Scientific American's* Nonsense," (Montgomery, Alabama: Apologetics Press, September, 2002) 31

4---Charles Darwin, Letter to J.D. Hooker [1 February] 1871, in Darwin, F., ed., *The Life and Letters of Charles Darwin,* [1898], (New York: Basic Books, Vol. II, 1959), 202-203.

5---Charles Darwin, *Descent of Man,* Vol. II, 389.

6---Richard Hutton, "Evolution: The Series," *WashingtonPost.com, Live Online*, Wednesday September 28, 2001.

7---Charles Darwin, *Origin of Species,* 637.

8---__________, Letter to J.D. Hooker [1 February] 1871, in Darwin, F., ed., *The Life and Letters of Charles Darwin,* [1898], (New York: Basic Books, Vol. II, 1959), 202-203.

9---Hubert P. Yockey, *Information Theory and Molecular Biology* (Cambridge: University Press, 1992), 257, quoted by Overman, 61, 62.

10---Michael Denton, *Evolution: A Theory in Crisis,*) 250, 264.

11---Jonathan Sarfati, "The Second Law of Thermodynamics: Answers to Critics," answers In Genesis.org/docs/370.asp#crystals (2002b), cited by Burt Thompson & Brad Harrub, "15 Answers to John Rennie and *Scientific American's Nonsense*" (Montgomery, Alabama: Apologetics Press, Inc., 2002) 44.

12---Dean L. Overman, *A Case Against Accident and Self-Organization,* 63, 64.

13---Percival Davis, Dean H. Kenyon, and Charles B. Thaxton, Academic Editor, *Of Pandas and People* (Dallas: Haughton Publishing Company, 1993) 2.

14---Dean L. Overman, *A Case Against Accident and Self-Organization*, 40, 41.

15---Michael Denton, *Evolution: A Theory in Crisis.* 296, 323.

16---Percival Davis, Dean H. Kenyon, and Charles B. Thaxton, 2, 3.

17---Ashby L. Camp, The Myth of Natural Origins (Tempe, Arizona: Ktisisa Publishing, 1994) 31, 32.

18---Dean L. Overman, *A Case Against Accident and Self-Organization*, 45, 46.

19---Carl Werner, "Criticisms of the Stanley Miller Experiment." *Evolution: the Grand Experiment* (Green Forest, Arkansas: New Leaf Press, 2007) 207.

20---Thaxton, C., Bradley, W., Olsen, R., *The Mystery of Life's Origin: Reassessing Current Theories* (Dallas: Lewis and Stanley Publishers, 1984) 76, 77, as cited by Carl Werner, *Evolution: The Grand Experiment*, 249.

21---__________., *The Mystery of Life's Origin: Reassessing Current Theories* (Dallas: Lewis and Stanley Publishers, 1984) 81, as cited by Carl Werner, *Evolution: The Grand Experiment*, 207.

22---Carl Werner, *Evolution: The Grand Experiment*, 207, referencing Thaxton, C., Bradley, W., Olsen, R., *The Mystery of Life's Origin: Reassessing Current Theories* (Dallas: Lewis and Stanley Publishers, 1984) 91. The reference to *"3.9 billion years"* is within the context of conventional time calibrations.

23---Hubert P. Yockey, *Information Theory and Molecular Biology* (Cambridge: Cambridge University Press, 1992), 235, 236, 238, and 335 as quoted by Overman, 48.

24---"'Intelligent Design' smacks of Creation by Another Name," *USA Today*, Editorial Page, August 9, 2005.

25---Gearld A. Kerkut, *Implications of Evolution* (London: Pergamon, 1960) 6, 7.

26---__________, *Implications of Evolution*, 157.

27---Sir Fred Hoyle, "The Big Bang in Astronomy," *New Scientist* (November 19, 1981) 92:527, cited by Harrub and Thompson, "Creationists Fight Back."

28---Sir Francis Crick, *Life Itself: Its Origin and Nature* (New York, Simon & Schuster, 1981) 88, cited by Brad Harrub and Bert Thompson, "Creationists Fight Back," (Montgomery, Alabama: Apologetics Press, 2002) 2.

29---Ian T. Taylor, *In the Minds of Men* (Minneapolis: TFE Publishing, 1996), 161 and 182, quoting Pasteur as cited by Rene J. Dubos, *Louis Pasteur: Freelance of Science* (New York: Charles Scribner's Sons, 1976) 395.

30---Roddy M. Bullock, "Darwinists on Design: Jumping to Confusions," citing 1860 Darwin letter to Asa Gray, a designist, *The ID Report*, www.Discover.org,, 2-28-2009.

31---Michael Denton, *Evolution: A Theory in Crisis*, 264.

32---__________, *Evolution: A Theory in Crisis.* 342.

33---__________, *Evolution: A Theory in Crisis*, 290, 291.

34---Mark Twain, *Life on the Mississippi* (Boston: J.R. Osgood, 1883), 156 as quoted by Brad Harrub, *Reason and Revelation*, May, 2001, 21(5):38.

Chapter V

1---Henry Gee, *In Search of Deep Time* ((New York: The Free Press, 1999) 116, 117.

2—Charles Darwin, *Origin*, 219.

3---Charles Darwin, cited by F. Darwin, *The Life and Letters of Charles Darwin,* (1881), Vol. 3, 309.

4---Charles Darwin, *On the Origin of Species by Means of Natural Selection, or the Preservation of Favored Races in the Struggle for Life,* (First Edition facsimile, 1859,

Cambridge: Harvard University Press) 184.

5---__________, *Descent,, Vol. I,* 207

6---__________, *Descent,* Vol. II, 389.

7---__________, *Descent,* Vol. 1, 206

8---James Gibson, letter to Warren L. Johns (August 28, 1997); citing David Raup, *Zoologic Record* published in *Paleobiology* 2 (1976) 279-288.

9---A. G. Fisher, Grolier Multimedia Encyclopedia, 1998, fossil section.

10---See *Darwin's Dilemma*, an Illustra Media DVD production, 2009.

11---Kathy Sawyer, "New Light on a Mysterious Epoch," *The Washington Post* (February 5, 1998); Copyright 1998, The *Washington Post.*

12---Francisco J. Ayala and James W. Valentine, *Evolving, The Theory and Processes of Organic Evolution*, 1979, 266.

13---Peter Ward & Donald Brownlee, *Rare Earth*, Feb 2000, 150.

14---T.S. Kemp, *Fossils and Evolution*, (Oxford University, Oxford University Press, 1999) 253.

15—Henry Gee, *In Search of Deep Time* ((New York: The Free Press, 1999) 32.

16---Ian T. Taylor, "The Ultimate Hoax: Archaeopteryx Lithographica," *Proceedings of the Second International Conference on Creationism, Vol II* (Pittsburgh: Creation Science Fellowship, Inc., 1990) 279-291.

17---Alan Feduccia, T. Lingham-Soliar and J.R. Hinchliffe, *Journal of Morphology* (2005) 266(2): 125-166 as reported by David Coppedge, "Have We Been Sold a Bill of Goods About Dinosaurs and BirdEvolution?", *Creation Matters* (St. Joseph, Missouri: Creation Research Society, September/October, 2005) 6, 7.

18---*National Geographic*, November, 1999.

19---Michael Denton, *Evolution: A Theory in Crises* (Bethesda, Maryland: Adler & Adler, 1986) 62, 358

20---Joseph Mastropaolo, "The Maximum-Power Stimulus Theory for Muscle," Creation Research Society Quarterly (St. Joseph, Missouri: Creation Research Society) Vol. 37, Number 4, March 2001, 213-219.

21---Darwin, *Origin,* 232.

22-- Henry Gee, *In Search of Deep Time,* 116, 117.

Chapter VI

1---Michael Denton, *Evolution: A Theory in Crisis*. 296, 323.

2---__________, *Evolution: A Theory in Crisis.* (Bethesda, Md.: Adler & Adler, 1986) 117, citing Darwin, C. (1858) in a letter to Asa Gray, 5 September, 1857, *Zoologist*, 16: 6297-99, see 6299.

3---__________, *Evolution: A Theory in Crisis*. 328, 329.

4---Charles Darwin letter to Asa Gray , cited by Adrian Desmond and James Moore, *Darwin,* (New York: W.W. Norton and Company, 1991) 456.

5---Desmond & Moore, *Darwin*, 475

6---Michael Denton, *Evolution: A Theory in Crisis*. 290, 291.

7---I. L. Cohen, *Darwin Was Wrong* (Greenvale, New York: New Research Publications, Inc., 1984) 38, 39, referencing H.G. Wells, Julian S. Huxley, and G.P. Wells, *The Science of Life* (New York: The Literary Guild) 41-43.

8---Michael Denton, *Evolution: A Theory in Crisis*. 328, 329.

9---Duane Arthur Schmidt, *And God Created Darwin* (Fairfax, Virginia: Allegiance Press, 2001) 24, 25.

10---I. L. Cohen, *Darwin Was Wrong* (Greenvale, New York: New Research Publications, Inc., 1984), 205.

11---Alan Hayward, *Creation and Evolution.* (Minneapolis: Bethany House Publishers, 1995) 35; referencing F. Hoyle, *The Universe: Past and Present Reflections* (University College, Cardiff, 1981).

12---I. L. Cohen, *Darwin Was Wrong* (Greenvale, New York: New Research Publications, Inc., 1984) 38, 39, referencing H.G. Wells, Julian S. Huxley, and G.P. Wells, *The Science of Life* (New York: The Literary Guild) 41-43.

13--Thomas Woodward, *Doubts About Darwin* (Grand Rapids: Baker Books, 2003) 44. Woodward referenced Sydney Fox in his review of *Mystery of Life's Origins in Quarterly Review of Biology*, June, 1985.

14---Harold Coffin, *Origin by Design* (Hagerstown, Maryland: Review and Herald Publishing Association, 1983) 379.

15---Joshua Lederberg, "A View of Genetics," *Science* 131 (3396) 1960: pp. 269-280 cited by Harold Coffin, *Origin by Design*, 377, 378.

16---Geoffrey Simmons, M.D., *What Darwin Didn't Know* (Eugene, Oregon: Harvest House Publishers, 2004) 43, 53.

17---Lee Spetner, *Not by Chance*. (Brooklyn, New York: The Judaica Press, Inc., 1997). 31.

18---Michael Denton, *Evolution: A Theory in Crisis*. 329.

19---__________, *Evolution: A Theory in Crisis*. 320, 321.

20---__________, *Evolution: A Theory in Crisis*. 342.

21---Walt Brown, *In the Beginning: Compelling Evidence for Creation and the Flood* (Phoenix, Ariz.: Center for Scientific Creation, 1996) 11, 12.

22---Carl Werner, *Evolution: the Grand Experiment,* 207.

23---Walt Brown, *In the Beginning: Compelling Evidence for Creation and the Flood* (Phoenix, Ariz.: Center for Scientific Creation, 1996) 11, 12.

24---Alan Hayward, *Creation and Evolution.* (Minneapolis: Bethany House Publishers, 1995) 35; referencing F. Hoyle, *The Universe: Past and Present Reflections* (University College, Cardiff, 1981).

25---W. Wells, "Taking Life to Bits," *New Scientist,* 155(2095):30-33, 1997 as cited in "How Simple Can Life Be?", *AnswerwinGenesis.org*, 2009.

26---Geoffrey Simmons, M.D., *What Darwin Didn't Know* (Eugene, Oregon: Harvest House Publishers, 2004) 43, 53.

27---Michael Denton, *Evolution: A Theory in Crisis,* 315.

28---Kirk R. Johnson and Richard E. Stucky, *Prehistoric Journey* (Boulder, Colorado: Roberts Rinehart Publishers, 1995) 16.

29---Dennis Wagner, "Top Ten Darwin and Design Science News Stories for 2010," ARN.org, /top10, December 21, 2010.

30---Albert Fleischmann, University of Erlangen Zoologist. See John Fred Meldau, ed., *Witnesses Against Evolution* (Denver: Christian Victory Publishing, 1968) 13.

31---Joyce Kilmer, *Trees and Other Poems* (New York, George H. Doran Company, 1914).

Chapter VII

1---Carl Werner, *Evolution: the Grand Experiment* (Green Forest, Arkansas: New Leaf Press, 2007) 192, 193.

2--- Michael Denton, *Evolution: A Theory in Crisis,* 306.

3---__________, 345.

4---George Javor, *CreationDigest.com*, Summer Edition, 2002.

5---Michael Denton, *Evolution: A Theory in Crisis* (Bethesda, Md.: Adler & Adler, 1986) 331.

6--- I. L. Cohen, *Darwin Was Wrong.* (Greenvale, New York: New Research Publications, Inc., 1984) 40-42.

7---Carl Werner, *Evolution: the Grand Experiment* (Green Forest, Arkansas: New Leaf Press, 2007), 194.

8--- I. L. Cohen, *Darwin Was Wrong,* 209.

9---Sir Fred Hoyle, "The Big Bang in Astronomy," *New Scientist* (November 19,1981) 92:527, cited by Harrub and Thompson, "Creationists Fight Back."

10---This data is from an unidentified internet source. While believed to be true, its accuracy has not be verified independently.

11---Bill Gates, *The Road Ahead* (Boulder: Blue Penguin, 1996) 228; from "ID in PS Curricula," 1999.

12---Lee M. Spetner, . *Not by Chance*. Brooklyn, New York: The Judaica Press, Inc., 1997, 30.

13--- George Javor, "5,000 Years of Stasis," ("Genomic Science: 21st Century Threat to 19th Century Superstition," www.*CreationDigest.com*, Summer Edition, 2002).

14---I. L. Cohen, *Darwin Was Wrong*, 35, 54.

15--- Michael Denton, *Evolution: A Theory in Crisis,* 334.

16---Joel Achenbach, "The Origin of Life Through Chemistry," *National Geographic*, March, 2006, 31.

17---Dean L. Overman, *A Case Against Accident and Self Organization* (New York: Rowman & Littlefield Publishers Inc., 1997) 59, citing Hoyle and Wickramasinge, 148, 24, 150, 30, and 31 as quoted in Charles B. Thaxton, Walter L. Bradley, and Roger Olsen, *The Mystery of Life's Origin: Reassuring Current Theories* (New York: Philosophical Library, 1984) 196.

18---" 'Intelligent Design' Smacks of Creation by Another Name," *USA Today,* Editorial Page, August 9, 2005.

19---Carl Werner, *Evolution: the Grand Experiment,* 195 citing Thaxton, C., Bradley, W., Olsen, R.., *The Mystery of Life's Origin: Reassessing Curent Theories* (Dallas: Lewis and StanleyPublishers,1984) 163, 164.

20---________, 196-7.

21---Jonathan Sarfati, "The Second Law of Thermodynamics: Answers to Critics," *Answers In Genesis,* (2002b), as cited by Burt Thompson and Brad Harrub, "15 Answers to John Rennie and *Scientific American's Nonsense*" (Montgomery, Alabama: Apologetics Press, Inc., 2002) 44. See also, A. Goffau, "Life With 482 Genes,"*Science.* 270 (5235), 445-6, 1995.

22---Lee M. Spetner, *Not by Chance* (Brooklyn, New York: The Judaica Press, Inc., 1997) 30.

23--- I. L. Cohen, *Darwin Was Wrong.*, 205

22---Michael Denton, *Evolution: A Theory in Crisis.* 342.

24---Michael Denton, *Evolution: A Theory in Crisis,,* 338.

25---I. L. Cohen, *Darwin Was Wrong.*, 40-42.

26---George Javor, "5,000 Years of Stasis," ("Genomic Science: 21st Century Threat to 19th Century Superstition," www.*CreationDigest.com*, Summer Edition, 2002).

27---Michael Denton, *Evolution: A Theory in Crisis,* 348, 357.

28---Richard Milton, *Shattering the Myths of Darwinism* (Rochester, Vt.: Park Street Press, 1997) 184; referencing Denton, *Evolution: A Theory in Crises.*

29--- I. L. Cohen, *Darwin Was Wrong,* 208.

30--- Michael Denton, *Evolution: A Theory in Crisis,* 324.

Chapter VIII

1--- Pierre-Paul Grassé, *The Evolution of Living Organisms,* (New York: Academic Press, 1977) 88, 103.

2---*The American Heritage Dictionary of the English Language,* Third Edition (New York, Houghton Mifflin Company, 1992) 754.

3--- Jobe Martin, *The Evolution of a Creationist* (Rockwall, Texas: Biblical Discipleship Publishers, 2002) 131, 132.

4---Ian Taylor, *In the Minds of Men*, 160, 161.

5--- Byron C. Nelson, *After Its Kind* (Minneapolis, Augsburg Publishing 1927) 98, 99.

6---Byron C. Nelson, *After Its Kind*, 101, referencing Alfred Russel Wallace, *Letters and Reminiscences*, 340.

7-- Jonathan Wells, *Icons of Evolution* (Washington, D.C.: Regnery Publishing, 2000) 180.

8--- Byron C. Nelson, *After Its Kind*, 101, quoting Wallace, *Letters and Reminiscences,* 95.

9---Alfred Russel Wallace, "The Present Position of Darwinism," *Contemporary Review,* August, 1908.

10--- Byron C. Nelson, *After Its Kind*, 101, citing *Theory of Evolution*, 163.

11---__________, 99,101 citing *Smithsonian Institute Report,* 1916, 343.

12---__________, 101, 102, citing *Science Progress*, January, 1925.

13---Albert Fleischmann, "The Doctrine of Organic Evolution in the Light of Modern Research," *Journal of the Transactions of the Victoria Institute* 65 (1933) 194-95, 205-6, 208-9.

14---A computerized comparison of the six editions can be checked out on Ben Fry's website, "The Preservation of Favored Traces."

15---Jonathan Wells, *Icons of Evolution*, 181.

16---Michael Denton, *Evolution: A Theory in Crisis* (Bethesda, Maryland: Adler & Adler, Publishers, inc., 1986) 75 citing Julian Huxley, *Evolution After Darwin* ed. Sol Tax, vol. 3 (Chicago: University of Chicago Press, 1960) 1-21, see 1.

17---See *A Scientific Dissent From Darwinism*, www.dissentfromdarwin.org.

18---Henry M. Morris, "What They Say," *Back to Genesis* (March 1999) a, b.

19---Lee Spetner, Not by Chance: Shattering the Modern Theory of Evolution (Brooklyn: Judaica Press, 1997) 160.

20--Bert Thompson, *The Scientific Case for Creation* (Montgomery, Alabama: Apologetic Press, Inc., 2002) 124.

21---Noble, et. al, *Parasitology,* sixth edition, "Evolution of Parasitism," Lea and Febiger, 1989, 516, as cited by Frank Sherwin, *Origins Issues,* "Natural Selection's Role in the Real World."

22---Colin Patterson, "Cladistics," *The Listener* (1982) 106:390.

23---Søren Løvtrup, *Darwinism: The Refutation of a Myth* (London: Croom Helm,1987) 352.

24---George Javor, "5,000 Years of Stasis," ("Genomic Science: 21st Century Threat to 19th Century Superstition," www.*CreationDigest.com*, Summer Edition, 2002).

25--See SanjaGupta, *CNN Health,* "Girl's mother just had 'feeling' something was wrong," *9-21-2010).*

26---Army's Edgewood Chemical Biological Center at Aberdeen Proving Grounds, Maryland, www.plosone.org/home.action, October 7, 2010.

27---Internet, www.beyondbooks.com. "Bacteria and Viruses."

28---Internet, http://en.wikipedia.org/wikl/Epigenetics.

Chapter IX

1---Richard Milton, *Shattering the Myths of Darwinism* (Rochester, Vt.: Park Street Press, 1997) 169, 170.

2--- Richard Lewontin, *The Triple Helix* (Cambridge, Massachusetts, HarvardUniversity Press, 2000) 91.

3---Lee Spetner, *Not by Chance: Shattering the Modern Theory of Evolution*, 97-103.

4 ---Byron C. Nelson, *After Its Kind*, 101, referencing Alfred Russel Wallace, *Letters and Reminiscences*, 340.

5—Lee Spetner, *Not by Chance: Shattering the Modern Theory of Evolution*, 139, 141.

6---Kevin Anderson, "Radio Interview with Dr. Kevin Anderson," *Creation Matters*, No. 4 July/August 2004, 1.

7---Lee Spetner, *Not by Chance: Shattering the Modern Theory of Evolution*, 131, 141, and 143.

8—Lee Spettner, *Not by Chance: Shattering the Modern Theory of Evolution*, 181, 198.

8---Michael Denton, *Evolution: A Theory in Crisis,* 334, 322.

10---Kevin Anderson, "Definition of Evolution," Anderson@nsric.ars.usda.gov, 9-4-2002, 1.

11---Marcia Barinaga, "Tracking Down Mutations That Can Stop the Heart ," *Science* 281 (July 3, 1998) 32.

12---Rick Weiss, "Defect Tied to Doubling of Risk for Colon Cancer," *The Washington Post*, August 26, 1997.

13---Tim Friend, "Gene Defect is Linked to Parkinson's," *USA Today* (June 27, 1997) and *USA Today*, January 17 (180, 2005.

14---Josie Glausiusz, "The Genes of 1996," *Discover* (January 1997) 36.

15---David A. Demick, "The Blind Gunman," *Impact* (El Cajon, Calif.: Institute for Creation Research, February, 1999) iv.

16--*The Star*, Ventura, California, June 24, 1997.

17---___________, June 24, 1997.

18---Reuters, "Genetic Error Causes Rapid-Aging Syndrome," *The Washington Post*, Thursday, April 17, 2003, A6.

19---Elizabeth Pennsi, "New Gene Found for Inherited Macular Degeneration," *Science* 281 (July 3, 1998) 31.

20---Daniel C. Weaver, "The River of Life," *Discover* (November 1997) 55.

21---Karen P. Steel and Steve D. M. Brown, "More Deafness Genes." *Science* 280 (May 29, 1998) 1403.

22---Rob Stein, "Sex May Rid Us of DNA Flaws," *The Washington Post* (February 1, 1999) A9.

23---See David Brown, "Scientists Discover 3 More Genes With Links to Alzheimer's Disease," *The Washington Post*, September 7, 2009, A3.

24---Mitch Lipka, *Consumer Ally,* "Meat Tainted With Deadly Virus is Being Sold to Consumers," September 28, 2010.

25---Liz Szabo, "Report: Just One Cigarette is Bad," *USA Today*, December 9, 2010, A-1.

26---Sean Pitman, *Detecting Design*.com, November 20, 2010.

27---Louis Bounoure *The Advocate*, 8 March 1984 , 17, quoted in *The Revised Quote Book*, 5. Bounoure has served as director of the Strasbourg Zoological Museum, and research director at the French National Center of Scientific Research.

28-Stephen J. Gould, Speech at Hobart College, February 14, 1980, cited by Luther Sunderland, *Darwin's Enigma* (El Cajon, California: Master Books, 1984), 106 (emphasis in original) cited by Bert Thompson and Brad Harrub, "*National Geographic* Shoots Itself in the Foot Again," (ApologeticsPress.Org online report, 2004) 36.

Chapter X

1---Michael J. Behe, *Darwin's Black Box.* (New York: The Free Press, 1996) 39.

2---___________, *Darwin's Black Box*, 15.

3---___________, *Darwin's Black Box*, 69-73.

4---___________, *Darwin's Black Box*, 31-36.

5---___________, *Darwin's Black Box,* 73.

6---___________, *Darwin's Black Box*, 79.

7---___________, *Darwin's Black Box*, 74-97.

8---___________, *Darwin's Black Box*, 86, 87.

9---___________, *Darwin's Black Box*, 90.

10---___________, *Darwin's Black Box*, 93, 94.

11---___________, *Darwin's Black Box*, 96, 97.

12---Michael Denton, "An Interview with Michael Denton," Access Research Network, Origins Research Archives, Vol. 15, Number 2, July 20, 1995.

13---___________, *Evolution: A Theory in Crisis.* (Bethesda, Md.: Adler & Adler, 1986) 117, citing Darwin, C. (1858) in a letter to Asa Gray, 5 September, 1857, *Zoologist*, 16: 6297-99, see 6299.

14--Donald R. Moeller, "Does a Smile Need 500 Million Years to Evolve?", *Creation Digest.com*, Spring Edition, 2002.

15--Albert Fleischmann, "The Doctrine of Organic Evolution in the Light of Modern Research," *Journal of the Transactions of the Victoria Institute 65* (1933) 194-95, 205-6, 208-9.

16---Frank Sherwin, "Un-Bee-lievable Vision," *Acts & Facts* (El Cajon, California: Institute for Creation Research, Vol. 35, No. 2, February, 2006) 5.

17---Steve Austin, *Grand Canyon: Monument to Catastrophe* (El Cajon, California: Institute for Creation Research, 1994) 145 as cited by Frank Sherwin, "Un-Bee-lievable Vision," *Acts & Facts* (El Cajon, California: ICR, Vol. 35, No. 2, February, 2006) 5.

18---Sir William Dawson, *The Story of Earth and Man* (New York: Harper and Brothers, 1887) 317, 322, 330, 339.

19---Søren Løvtrup (Swedish biologist), Darwinism: The Refutation of a Myth (New York: Croom Helm, 1987) 422.

20---Dennis Wagner, "2009 Annual Report: The Key Darwin and Design Science News Stories of the Year," *Access Research Network,* December 30, 2009.

21---Ernst Mayer,*The Growth of Biological Thought: Diversity, Evolution and Inheritance* (Cambridge, Massachusetts: The Belknap Press of Harvard University Press, 1982) 584.

Chapter XI

1---*The Holy Bible,* New International Version, Genesis 1:20.

2---Charles Darwin, *Origin of Species*, 647, 453.

3---A. G. Fisher, *Grolier Multimedia Encyclopedia*, 1998, fossil section.

4---Peter Ward & Donald Brownlee, *Rare Earth*, Feb 2000, 150.

5---Henry Gee, *In Search of Deep Time* (London: Comstock Publishing Associates, 1999) 116, 117

6---Henry Gee, *In Search of Deep Time,* 108, 110.

7---Kathy Sawyer, "New Light on a Mysterious Epoch," *The Washington Post* (February 5, 1998); Copyright 1998, The *Washington Post.*

8---Charles Darwin, *Origin of Species,* 647.

9---Charles Q. Choi, "Cache in Chinese Mountain Reveals 20,000 Prehistoric Fossils," *LifeScienc.com,* December 22, 2010.

10--Darwin, C. (1881) in Darwin, F., *The Life and Letters of Charles Darwin,* (London: John Murray, 1888) vol. 3, 248; cited by Michael Denton, *Evolution: A Theory in Crisis,* 163

11---Charles Darwin, *Origin,* 647.

12---See *Creation ex Nihlo*, December, 2000, 6.

13---Henry Gee, *In Search of Deep Time,*108, 110.

14---Charles Darwin, *Origin*, 647, 453.

15---Erik Stokstad, *Science,* 5 December 2003, 1645.

16---*Creation Matters* (Vol. 15, No. 6, November/December, 2010) 5, citing Kent State University (2010, November 9) as reported by www.physorg.com/news/2010-11.

17---Michael Denton, *Evolution: A Theory in Crisis.* (Bethesda, Md.: Adler & Adler, 1986) 298, 302.

18---Henry Gee, *In Search of Deep Time,* 133.

19---Steven A. Austin, ed., *Grand Canyon: Monument to Catastrophe* (Santee, California; Institute for Creation Research, 1994) 149.

20---Peter D. Ward, "Coils of Time," *Discover* (March 1998) 106.

21---Michael Denton, *Evolution: A Theory in Crisis.* (Bethesda, Md.: Adler & Adler,1986)
    298, 302.

22--David Coppedge, "Speaking of Science," *Creation Matters,* January/February,
    2010, 5.

23---Murray Wardrop, "Scientists Draw Squid Using its 150-million-year-old Fossilized Ink,"
    Telegraph.co.UK, 19 August, 2009.

24---Charles Darwin, *Origin*, 219.

25---__________, *Origin*, 617, 618.

26---__________, *Origin*, 406.

27---__________, *Origin,* 232.

28---Henry Gee, *In Search of Deep Time*, 177, 155.

29---__________, *In Search of Deep Time*, 127.

30---Ernst Mayr, *The Growth of Biological Thought: Diversity, Evolution and Inheritance*
    (Cambridge, Massachusetts: The Belknap Press of Harvard University Press, 1982) 524.

31---Charles Darwin, *Origin*, 406.

32---R.L. Wysong, The Creation-Evolution Controversy. (Midland, Michigan: Inquiry Press,
    1978) 348, 352-354.

33---Genesis 1:21, *The Holy Bible, New International Version.*

34---Lyall Watson, "The Water People," *Science Digest*, 90[5]:44, May, as cited by Eric Lyons
    and Kyle Butt, *The Dinosaur Delusion*, (Montgomery, Alabama: Apologetics Press, Inc.,
    2008) 140.

© Warren L. Johns

**One-thousand replicas of the most colorful illuminated copies
of the Gutenbert Bible were published in 1961.**

Achenbach, Joel, "When Yellowstone Explodes," *National Geographic,* August, 2009.

Achenbach, Joel, "The Origin of Life Through Chemistry," *National Geographic*, March, 2006, 31.

Anderson, Kevin, "Radio Interview with Dr. Kevin Anderson," *Creation Matters*, No. 4 July/August 2004.

__________, "Definition of Evolution," Anderson@nsric.ars.usda.gov, 2002.

Austin, Steven A. "Excess Argon With Mineral Concentrates From the New Dacite Lava Dome at Mount St. Helens Volcano." *Creation Ex Nihilo Technical Journal,* vol. 10, part 3, 1996; as reported in "Acts and Facts," *Institute for Creation Research*, May, 997.

__________, Editor. *Grand Canyon: Monument to Catastrophe,* .Santee, Calif.: Institute for Creation Research, 1994.

__________, "Interpreting Strata of Grand Canyon," in Austin, *Grand Canyon: Monument to Catastrophe*, Santee, California: Institute for Creation Research. 1994.

__________, *"Nautiloid Mass Kill and Burial Event, Redwell Limestone, Grand Canyon* Region," *Pittsburgh International Conference on Creationism, 2003.*

Ayala, Francisco J. and James W. Valentine, *Evolving, The Theory and Processes of Organic Evolution, 1979.*

Begley, Sharon, "How to Think Like a Scientist," *Newsweek,* July 9, 2007.

Behe, Michael J. *Darwin's Black Box,* New York: The Free Press, 1996.

Bird, W.R., *The Origin of Species Revisited, Vol. I* **, Nashville: Regency, 1991.**

Bounoure, Louis, *The Advocate,* 8 March 1984, quoted in *The Revised Quote Book.*

Brown, Anthony, *The Times,* London, UK, 20 February 2003, reported by *info@creationresearch.net,* February 27, 2003.

Brown, David, "Limits to Genetic Evolution," *The Washington Post*, July 7, 2003.

__________, **"**Scientists Discover 3 More Genes With Links to Alzheimer's Disease," *The Washington Post*, September 7, 2009, A3.

Brown, Walt, *In the Beginning: Compelling Evidence for Creation and the Flood* Phoenix, Ariz.: Center for Scientific Creation, 1996.

Bullock, Roddy M., "Darwinists on Design: Jumping to Confusions," citing 1860 Darwin letter to Asa Gray, a designist, *The ID Report*, www.Discover.org, February 28, 2009.

Camp, Ashby L., *The Myth of Natural Origins*, Tempe, Arizona: Ktisis Publishing, 1994 Edition.

Choi, Charles Q., "Cache in Chinese Mountain Reveals 20,000 Prehistoric Fossils," *LifeScienc.com*, December 22, 2010.

Coffin, Harold, with Robert H. Brown, *Origin by Design*. Hagerstown, Md.: Review and Herald Publishing Association, 1983.

Cohen, I. L., *Darwin Was Wrong*. Greenvale, New York: New Research Publications, Inc., 1984.

Coles, Peter, "Boomtime," *New Scientist*, March 3, 2007.

Cooper-Cole, Fay, *The Scopes Trial* , Birmingham, Ala.: The Legal Classics Library, 1984, reprint of *The World's Most Famous Trial* , Cincinnati: National Book Co., 1925.

Coppedge, David, "Speaking of Science," *Creation Matters,* January/February, 2010.

Cornelius, Richard M., "Scopes Trial: The Trial Gavel Heard Round the World," as excerpted from *Impact* (Dayton, Tennessee: Bryan College, 2000, v-xiii.

Creation Matters, Vol. 15, No. 6, November/December, 2010.

Crick, Sir Francis, *Life Itself*, New York: Simon Schuster, 1981.

Culotta . Elizabeth and Elizabeth Pennisi, "Evolution in Action," Science, 23 December 2005, Vol. 310, 1878-79.

Darwin, Charles Robert, *On the Origin of Species by Means of Natural Selection, or the Preservation of Favored Races in the Struggle for Life*, First Edition facsimile, 1859, Cambridge: Harvard University Press.

__________, *The Origin of Species* (Sixth Edition), New York: Random House, Inc., 1993.

__________, *The Descent of Man, and Selection in Relation to* Sex, Vol. I & Vol. II Princeton, N.J., Princeton University Press, 1981.

__________, Letter to Asa Gray, cited by Adrian Desmond and James Moore, *Darwin*, New York: W.W. Norton and Company, 1991.

__________, Letter to J.D. Hooker [1 February] 1871, in Darwin, F., ed., *The Life and Letters of Charles Darwin*, [1898], New York: Basic Books, Vol. II, 1959.

__________, (1881) in Darwin, F., *The Life and Letters of Charles Darwin*, London: John Murray, 1888, Vl. 3, cited by Michael Denton, *Evolution: A Theory in* Crisis.

Davies, Paul, *The Cosmic Blueprint: New Discoveries in Nature's Ability to Order the Universe*, New York: Simon & Schuster, 1988.

__________, *The Mind of God*, New York: Simon & Schuster, 1992.

Davis, Percival & Dean H. Kenyon, Charles B. Thaxton, Academic Editor, *Of Pandas and People*, Dallas: Haughton Publishing Company, 1993.

Desmond, Adrian & James Moore, *Darwin.* New York, Warner Books, Inc., 1991.

Denton, Michael, *Evolution: A Theory in Crisis.* Bethesda, Md.: Adler & Adler, 1986.

__________, "An Interview with Michael Denton," Access Research Network, Origins Research Archives, Vol. 15, Number 2, July 20, 1995.

Dawson, Sir William, *The Story of Earth and Man* (New York: Harper and Brothers, 1887).

Ellegren, Hans, "Genomics: The Dog Has Its Day," *Nature*, 8 December 2005, 745, referenced in "Scientists Put Dogs on Genome Map," *USA Today* ,Thursday, December 8, 2005.

Feduccia, Alan, as quoted by Kathy A. Svitil, "Discover Dialogue," *Discover*, February, 2002.

__________, **Feduccia, Alan, T. Lingham-Soliar and J.R. Hinchliffe,** *Journal of Morphology*, 2005, 266(2).

Fisher, A. G., *Grolier Multimedia Encyclopedia*, 1998.

Fleischman, Albert, "The Doctrine of Organic Evolution in the Light of Modern Research," *Journal of the Transactions of the Victoria Institute* 65, 1933.

Gates, Bill, *The Road Ahead* (Boulder: Blue Penguin, 1996) 228; from "ID in PS Curricula," 1999.

Gibson, James, Letter to Warren L. Johns. August 28, 1997.

Gould, Stephen Jay, *The Panda's Thumb*, New York: W.W. Norton, 1980.

__________, "The Return of Hopeful Monsters, "*Natural History,* 86[4]:22-30, June-July, 1977.

__________, Speech at Hobart College, February 14, 1980, cited by Luther Sunderland, *Darwin's Enigma*, El Cajon, California: Master Books, 1984 cited by Bert Thompson and Brad Harrub, "*National Geographic* Shoots Itself in the Foot Again," Apologetics Press.Org online report, 2004.

Grassé, Pierre-Paul, *The Evolution of Living Organisms*,, New York:Academic Press, 1977.

Handel , George Frederic, *The Messiah* (1741-42).

Hayward, Alan, *Creation and Evolution.* (Minneapolis: Bethany House Publishers, 1995), referencing F. Hoyle, *The Universe: Past and Present Reflections* (University College, Cardiff, 1981).

Hayden, Thomas, "A Theory Evolves," *U.S. News & World Report*, July 29, 2002..

Hayward, Alan. *Creation and Evolution.* Minneapolis: Bethany House Publishers, 1995.

Higgins, Adrian, "Why the Red Delicious No Longer Is," *The Washington Post National Weekly Edition*, August 15-21, 2005.

Holy Bible, King James Version, New York: Oxford University Press; also New International Version, Grand Rapids: Zondervan Bible Publishers, 1978.

Hoyle, Sir Fred and Chandra Wickramasinghe, *Evolution from Space,* London: J.M. Dent & Sons, 1981.

Hunter, George H., *Civic Biology* , 1914.

Hutton, Richard, "Evolution: The Series," *Washington Post.com, Live Online,* Wednesday, September 28, 2001.

Javor, George, "5,000 Years of Stasis," "Genomic Science: 21st Century Threat to 19t Century Superstition," www.*CreationDigest..com,* Summer Edition, 2002.

Johns, Warren LeRoi. *Ride to Glory,* Brookeville, Maryland: General Title, Inc., 1999.

__________, *Beyond Forever,* Smithville, Tennessee: *Creation Digest,* 2007.

__________, *Genesis File,* Smithville, Tennessee, www.GenesisFile.clom, 2010.

Johnson, Kirk R. and **Richard K. Stucky** *Prehistoric Journey,* Boulder, Colorado: Roberts Rinehart Publishers, 1995.

Kerkut, G.A., *Implications of Evolution,* New York: Pergamon Press, 1965.

Kilmer, Joyce, *Trees and Other Poems* (New York, George H. Doran Company, 1914).

Kime, Wesley, internationally recognized portrait artist, writes with the academic credentials of a Physician, finished at the top of his 1953 School of Medicine class, 2008 statement to Editor..

Lederberg Joshua, "A View of Genetics," *Science* 131 (3396) 1960: pp. 269-280 cited by Harold Coffin, *Origin by Design.*

Lester, Lane P., and **Raymond G. Bohlin.** *The Natural Limits to Biological Change* Dallas, Texas: Probe Books, 1989.

Lewontin, Richard , *The Triple Helix,* Cambridge, MS, Harvard University Press, 2000.

Løvtrup, Søren. *The Refutation of a Myth,* New York: Croom Helm, 1987.

Macbeth, Norman. *Darwin Retried: An Appeal to Reason.* Boston: The Harvard Common Press, 1978.

Martin, Jobe, *The Evolution of a Creationist* , Rockwall, Texas: Biblical Discipleship Publishers, 1994.

Mastropaolo, Joseph, "The Maximum-Power Stimulus Theory for Muscle, "Creation Research Society Quarterly (St. Joseph, Missouri: Creation, Research Society) Vol. 37, No. 4, March 2001.

Mayr, Ernst. *Systematics and the Origin of Species.* New York: Columbia University Press, 1942; Dover Publications paperback, 1964.

__________, *The Growth of Biological Thought: Diversity, Evolution and Inheritance,* Cambridge, Massachusetts: The Belknap Press of Harvard University Press, 1982.

Metcalf, Maynard M., in Adams, Leslie B., Jr., Editor/Publisher, The Scopes Trial Birmingham, Alabama: The Legal Classics Library, 1984; a reprint of The World's Most Famous Trial, Third Edition, Cincinnati: National Book Company, 1925.

Milner, R., *The Encyclopedia of Evolution: Humanity's Search for Its Origins,* New York: Facts o File Publishers, 1990.

Milton, Richard. *Shattering the Myths of Darwinism.* Rochester, Vermont.: Park Street Press, 1997.

Moeller, Donald R., "Does a Smile Need 500 Million Years to Evolve?", *CreationDigest.com,* Spring Edition, 2002.

Morris, Henry M., "The Microwave of Evolution," *Back to Genesis,* August, 2001, a.

__________. "What They Say." *Back to Genesis.* N. Santee, Calif.: Institute for Creation Research, 1999.

National Geographic, November, 1999; and promo, "BODY: The Complete Human."

Nelson, Byron C. *After Its Kind.* Minneapolis: Augsburg Publishing House, 1927.

__________, referencing Alfred Russell Wallace, *Letters and Reminiscences.*

Oliver & Boyd, *Contemporary Botanical Thought,* 1971.

Overman, Dean L., *A Case Against Accident and Self-Organization.* New York: Rowman & Littlefield Publisher, Inc., 1997.

Patterson, Colin, "Can You Tell be Anything About Evolution," lecture as transcribed by Wayne Frair and reported in "Bridge to Nowhere?", *CreationDigest.com,* Autumn 2004 Edition.

Pitman, Sean, *Detecting* Design.com, November 20, 2010l..

Quammen, David, *The Reluctant Mr. Darwin,* New York: Atlas Books, W.W. Norton & Company, 2006.

Sarfati, Jonathan, *Refuting Evolution.* Green Forest, Arkansas: Master Books, 1999.

__________, "The Second Law of Thermodynamics: Answers to Critics," answersingenesis.org/docs/370.asp#crystals (2002b).

Sawyer, Kathy. "New Light on a Mysterious Epoch," *The Washington Post,* February 5, 1998.

Schmidt, Duane Arthur, *And God Created Darwin* (Fairfax,Virginia: Alliance Press, 2001.

Sherwin, Frank, "Un-Bee-lievable Vision," *Acts & Facts,* El Cajon, California: Institute for Creation Research, Vol. 35, No. 2, February, 2006.

__________, *Origins Issues,* "Natural Selection's Role in the Real World," citing Noble, et. al, *Parasitology,* sixth edition, "Evolution of Parasitism," Lea and Febiger, 1989.

Simmons, Geoffrey, M.D., *What Darwin Didn't Know*, Eugene, Oregon: Harvest HousePub., 2004.

Smith, Wolfgang, *Teilhardism and the New Religion*. Rockford, Ill.: Tan Books, 1988.

Spetner, Lee M. *Not by Chance*. Brooklyn, New York: The Judaica Press, Inc., 1997.

Stokstad, Erik, "Gutsy Fossil Record for Staying the Course," *Science*, Vol. 302, 5 December 2003.

Sunderland, Luther, *Darwin's Enigma: Fossils and Other Problems*, San Diego: Master Books, 1988.

Swarts, Mark, "Scientists Confirm Age of the Oldest Meteorite Collision on Earth, "*SpaceDaily.com*, August 23, 2002.

Szabo, Liz, "Report: Just One Cigarette is Bad," *USA Today*, December 9, 2010,.

Taylor, Ian T. *In the Minds of Men*. Minneapolis, Minn.: TFE Publishing, 1991.

__________, "The Ultimate Hoax: Archaeopteryx Lithographica," *Proceedings of the Second International Conference on Creationism, Vol.. II* , Pittsburgh: Creation Science Fellowship, Inc., 1990.

__________, "The Idea of Progress," *The Fifth International Conference on Creationism*, Pittsburgh: Creation Science Fellowship, Inc., 2003.

Thaxton, Charles B., **Walter L. Bradley** and **Roger L. Olsen**, *The Mystery of Life's Origin*, New York: Philosophical Library, 1984.

Thompson, Bert and **Brad Harrub**, "15 Answers to John Rennie and *Scientific American's* Nonsense." Montgomery, Alabama: Apologetics Press, 2002.

__________, "*National Geographic* Shoots Itself in the Foot Again," Apologetics Press.Org online r eport, 2004.

Twain, Mark, *Life on the Mississippi* (Boston: J.R. Osgood, 1883), 156 as quoted by Brad Harrub, *Reason and Revelation*, May, 2001, 21(5).

Wagner, Dennis, "2009 Annual Report: The Key Darwin and Design Science News Stories of the Year," *Access Research Network*,. December 30, 2009.

__________, "Top Ten Darwin and Design Science News Stories for 2010," ARN.org, /top10, December 21, 2010.

Wallace, Alfred Russell. *The Geographical Distribution of Animals, With a Study of the Relations of the Living and Extinct Faunas as Elucidating Past Changes of the Earth's Surface.* New York: Harper, 1876.

__________, *Darwinism*, London and New York, Macmillan and Co., 1890.

__________, "The Present Position of Darwinism," *Contemporary Review*, August, 1908.

Ward, Peter D., "Coils of Time," *Discover*, March 1998.

Ward, Peter & Donald Brownlee, *Rare Earth*, Feb 2000.

Wardrop, Murray, "Scientists Draw Squid Using its 150-million-year-old Fossilized Ink," *Telegraph.co.UK*, 19 August, 2009.

Watson, Lyall, "The Water People," *Science Digest*, 90[5]:44, May, as cited by Eric Lyons and Kyle Butt, *The Dinosaur Delusion*, Montgomery, Alabama: Apologetics Press, Inc., 2008.

Wells, Jonathan, *Icons of Evolution*. Washington, D.C.: Regnery Publishing, Inc., 2000.

__________, "Survival of the Fittest," *The American Spectator*, December 2000/January 2001.

Wells.W., "Taking Life to Bits," *New Scientist*, 155(2095):30-33, 1997 as cited in "How Simple Can Life Be?", *AnswerwinGenesis.org*, 2009.

Werner, Carl, *Evolution: The Grand Experiment*, Green Forest, Arkansas: New Leaf Press, 2007

Woodward, Thomas, *Doubts About Darwin*. Grand Rapids: Baker Books, 2003.

Wysong, Randy L., *The Creation-Evolution Controversy*East Lansing, Michigan: Inquiry Press, 1976.

Yockey, Hubert P., *Information Theory and Molecular Biology*. Cambridge: Cambridge University Press, 1992.

Zimmer, Carl, "Testing Darwin," *Discover*, February, 2005.

__________, "A New Step in Evolution," internet *Seed Design Series*, posted June 2, 2008.

Studio Six, 5-18-2006

Author
Warren LeRoi Johns

Johns practiced law as a career in California, Maryland, and the District of Columbia until partial retirement in the summer of 1992. Admitted to practice before the United States Supreme Court in 1963, he has been a member of the American Association for the Advancement of Science.

His non-fiction *Dateline Sunday, U.S.A.,* drew national attention as a legal history documenting blue law confrontation with the U.S. Constitution's first amendment. His 1999 *Ride to Glory* targeted some of evolution's more obvious shortfalls while a lawyer's academic perspective documented evolution's most obvious *"flaws"* and *"holes"* in his 2007 *Beyond Forever.*

A 1958 graduate of the University of Southern California's Law Center, and holder of La Sierra University's 1994 "Alumnus of the Year" award, the author's professional resume appears in *Who's Who in American Law, Who's Who in America,* and *Who's Who in the World.*

Warren L. Johns, Esq. (ret.)
wj1935@yahoo.com

"...The fool says in his heart, 'There is no God.' "
Psalms 14:1

157

CPSIA information can be obtained
at www.ICGtesting.com
Printed in the USA
237755LV00005BA